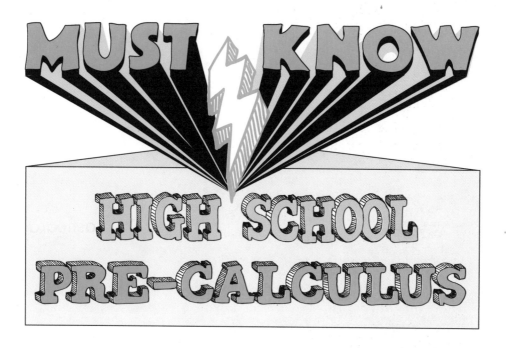

MUST KNOW

HIGH SCHOOL PRE-CALCULUS

Chris Monahan

McGraw Hill

New York Chicago San Francisco Athens London Madrid
Mexico City Milan New Delhi Singapore Sydney Toronto

2 3 4 5 6 7 8 9   LCR   24 23

ISBN   978-1-260-45815-2
MHID   1-260-45815-6

e-ISBN   978-1-260-45813-8
e-MHID   1-260-45813-X

Interior design by Steve Straus of Think Book Works.
Cover and letter art by Kate Rutter.

McGraw-Hill Education books are available at special quantity discounts to use as premiums and sales promotions or for use in corporate training programs. To contact a representative, please visit the Contact Us pages at www.mhprofessional.com.

This book is dedicated to my four children – Kate, Brendon, Andrew, and Kristen. Thank you for all your support and encouragement all these years.

Contents

12 Polar and Parametric Equations 289

13 Complex Numbers 335

14 Introduction to Calculus 348

Introduction

Welcome to your new pre-calculus book! Let us try to explain why we believe you've made the right choice. This probably isn't your first rodeo with either a textbook or other kind of guide to a school subject. You've probably had your fill of books asking you to memorize lots of terms (such is school). This book isn't going to do that—although you're welcome to memorize anything you take an interest in. You may also have found that a lot of books jump the gun and make a lot of promises about all the things you'll be able to accomplish by the time you reach the end of a given chapter. In the process, those books can make you feel as though you missed out on the building blocks that you actually need to master those goals.

With *Must Know High School Pre-Calculus*, we've taken a different approach. When you start a new chapter, right off the bat you will immediately see one or more **must know** ideas. These are the essential concepts behind what you are going to study, and they will form the foundation of what you will learn throughout the chapter. With these **must know** ideas, you will have what you need to hold it together as you study, and they will be your guide as you make your way through each chapter.

To build on this foundation, you will find easy-to-follow discussions of the topic at hand, accompanied by comprehensive examples that show you how to apply what you're learning to solving typical pre-calculus questions. Each chapter ends with review questions—more than 250 throughout the book—designed to instill confidence as you practice your new skills.

This book has other features that will help you on this pre-calculus journey of yours. It has a number of sidebars that will both help provide helpful information and serve as a quick break from your studies. The **BTW** sidebars ("by the way") point out important information, as well as tell you what to be careful about pre-calculus-wise. Every once in a while, an 🌐**IRL** sidebar ("in real life") will tell you what you're studying has to do with the real world; other IRLs may just be interesting factoids.

In addition, this book is accompanied by a flashcard app that will give you the ability to test yourself at any time. The app includes more than 100 "flashcards" with a review question on one "side" and the answer on the other. You can either work through the flashcards by themselves or use them alongside the book. To find out where to get the app and how to use it, go to the next section, The Flashcard App.

We also wanted to introduce you to your guide throughout this book. Chris Monahan has more than 30 years' experience teaching math at the high school and college levels. He knows what you should get out of a pre-calculus course, and his strategies will help get you there. Chris has seen the typical kinds of problems that students can have with pre-calculus, and he is experienced at solving those difficulties. In this book, he uses that experience to show you not only the most effective way to learn a given concept but how to get yourself out of traps you might have fallen into. We've had the pleasure of working with Chris before and are confident that we're leaving you in good hands.

Before we leave you to Chris's surefooted guidance, let us give you one piece of advice. While we know that saying something "is the *worst*" is a cliché, if anything *is* the worst in pre-calculus, it's getting used to functions. Let Chris introduce you to the concept and show you how to apply it confidently to your pre-calculus work. Take our word for it, mastering functions will leave you in good stead for the rest of your math career.

Good luck with your studies!

The Editors at McGraw-Hill

The Flashcard App

This book features a bonus flashcard app. It will help you test yourself on what you've learned as you make your way through the book (or in and out). It includes 100-plus "flashcards," both "front" and "back." It gives you two options as to how to use it. You can jump right into the app and start from any point that you want. Or you can take advantage of the handy QR Codes near the end of each chapter in the book; they will take you directly to the flashcards related to what you're studying at the moment.

To take advantage of this bonus feature, follow these easy steps:

Search for **Must Know High School** App from either Google Play or the App Store.

↓

Download the app to your smartphone or tablet.

↓

Once you've got the app, you can use it in either of two ways.

↙ ↘

Just open the app and you're ready to go.	Use your phone's QR Code reader to scan any of the book's QR Codes.
You can start at the beginning, or select any of the chapters listed.	You'll be taken directly to the flashcards that match your chapter of choice.

↘ ↙

Get ready to test your pre-calculus knowledge!

 # Functions

athematics is known for its ability to convey a great deal of information with the usage of the minimum number of symbols. While this may be initially confusing (if not frustrating) for the learner, the notation of mathematics is a universal language. In this chapter, you will learn about function notation.

Relations and Functions

One of the major concepts used in mathematics is relations. A **relation** is any set of ordered pairs. The set of all first elements (the input values) is called the *domain*, while the set of second elements (the output values) is called the *range*. Relations are traditionally named with a capital letter. For example, given the relation

$$A = \{(2, 3), (-1, 5), (4, -3), (2, 0), (-9, 1)\}$$

the domain of A (written DA) is $\{-9, -1, 2, 4\}$. The domain was written in increasing order for the convenience of reading, but this is not required. The element 2, which appears as the input for two different ordered pairs, needs to be written just the one time in the domain. The range of A (written RA) is $\{-3, 0, 1, 3, 5\}$.

The inverse of a relation is found by interchanging the input and output values. For example, the inverse of A (written A^{-1}) is

$$A^{-1} = \{(3, 2), (5, -1), (-3, 4), (0, 2), (1, -9)\}.$$

Do you see that the domain of the inverse of A is the same set as the range of A and that the range of the inverse of A is the same as the domain of A? This is very important.

▶ Determine the domain and range of the relation:

$$B = \{(3, -2), (-4, 5), (2, 1), (-5, -8), (-3, -2), (9, 7)\}$$

▶ The domain is $\{-5, -4, -3, 2, 3, 9\}$, while the range is $\{-2, 5, 1, -8, -2, 7\}$.

Functions are one of the most important concepts in all of mathematics. To understand this, let's first look at exactly what a function is. A **function** is a relation in which for each first element in the relation there is a unique second element. (To restate this: for each element in the domain there is a unique element in the range.) If a relation is a function, then there can only be one answer for each input value.

You can see that in the previous example A is not a function because 2 is used twice as an input with different output values, while A^{-1} is a function.

Verbal descriptions of examples identifying functions require us to give thought to all the possibilities of what is included. A graphical representation covers all possibilities.

EXAMPLE

▶ Is the relation represented by the graph below a function?

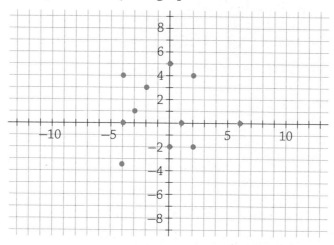

▶ No, it is not because some input values (−4 and 2) have multiple output values.

A very useful guideline for determining functions is the **vertical line test.** If a vertical line passes through multiple points anywhere on a graph, then the relation determined by those points is not a function.

  Let's apply the vertical line test to a graph. Does this graphical representation represent a function?

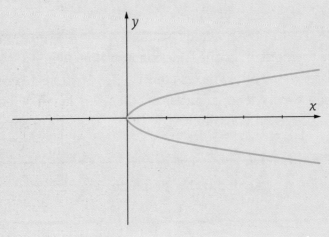

The graph does not satisfy the vertical line test, so it does not represent a function.

As you well know, mathematics uses a lot of notation. If you wish, you can think of it as codes or shorthand. With functions, we get the ability to work with multiple equations and have them clearly identified. We can discuss the functions $f(x) = 4x - 7$, $g(x) = 8x^2 - 5x - 3$, and $p(x) = \dfrac{2x + 3}{3x - 2}$ and simply refer to them as $f(x)$, $g(x)$, or $p(x)$ and everyone will know which algebraic rule we are using.

At this point in your study of mathematics, you should know that the function notation is a substitution process. With $f(x) = 4x - 7$, $f(-3) = 4(-3) - 7 = -19$, while $f(2t) = 4(2t) - 7 = 8t - 7$.

IRL Many disciplines use functions and they do call them $f(x)$. Economists might use $R(p)$ to represent the revenue earned when the price of an item sold is $\$p$. A chemist might use the notation $V(t)$ to represent the volume of a gas when the temperature of the gas is t degrees. It is commonplace to use $f(x)$ when first learning about functions because f is the first letter of function and x is frequently used to represent the input variable.

The domain of a function is the set of values for which the function is defined. When the function in question is a generic algebraic function, we realize that the domain will be the set of real numbers *except* for those input values that will result in an attempt to divide by 0 or taking an even root (such as the square root or fourth root) of a negative number.

EXAMPLE

▶ Determine the domain of the function $k(x) = \dfrac{\sqrt{2x-5}}{x-7}$.

▶ We know that $x = 7$ cannot be allowed because it causes the denominator of the fraction to equal 0. We also know that $2x - 5 \geq 0$, so the radicand (the number inside the radical) is positive. This results in $x \geq \dfrac{5}{2}$.

▶ Therefore, the domain of $k(x)$ is $x \geq \dfrac{5}{2}$ and $x \neq 7$.

Finding the range of a function is not as easy to do, and you will therefore see that this subject is examined a number of times in this book.

The arithmetic of functions is just a fancy name for doing the algebraic arithmetic you have been doing since you began to study Algebra I.

EXAMPLE

▶ If $f(x) = 8x^2 - 16x$ and $q(x) = \dfrac{5x+9}{x+4}$, determine the value of $\dfrac{f(3)}{q(3)}$. (This problem could also have been written as $\dfrac{f}{q}(3)$).

▶ With $f(x) = 8x(x-2)$ so $f(3) = 8(3)(3-2) = 24$, while $q(3) = \dfrac{5(3)+9}{3+4} = \dfrac{24}{7}$.

▶ Therefore, $\dfrac{f(3)}{q(3)} = \dfrac{24}{\dfrac{24}{7}} = 24 \times \dfrac{7}{24} = 7$.

Composition of Functions

Evaluating a function with the output of another function is called a **composition** of functions. Using $f(x) = 8x^2 - 16x$ and $q(x) = \dfrac{5x+9}{x+4}$, $q(-3) = \dfrac{5(-3)+9}{(-3)+4} = \dfrac{-6}{1} = -6$ and $f(-6) = 8(-6)^2 - 16(-6) = 8(36) - 96 = 288 - 96 = 192$. There are two ways to represent this action using function notation, $f(g(-3))$ or $f \circ g(-3)$. We can determine the rule for this composed function in terms of one variable:

$$f(g(x)) = -8(g(x))^2 - 16(g(x)) = -8\left(\frac{5x+9}{x+4}\right)^2 - 16\left(\frac{5x+9}{x+4}\right).$$

Of course, we can do as much algebra as we choose to simplify the expressions involved. It is critical to note that $g(f(x)) = \dfrac{5(f(x))+9}{f(x)+4} = \dfrac{5(8x^2-16x)+9}{8x^2-16x+4} = \dfrac{40x^2-80x+9}{8x^2-16x+4}$ is not the same as $f(g(x))$.

▶ The number of articles for a product sold by a company is given by the function $n(p) = -0.75p + 150$, where p represents the price, in dollars, to produce one unit. The profit, P, for selling n units of this product is $P(n) = -4n^2 + 20n - 25$. Express the profit in terms of the price to produce each article.

$$\begin{aligned}
P(n(p)) &= -4(-0.75p+150)^2 + 20(-0.75p+150) - 25 \\
&= -4(0.5625p^2 - 225p + 22{,}500)^2 - 15p + 3000 - 25 \\
&= -2.25p^2 + 900p - 90{,}000 - 15p + 2975 \\
&= -2.25p^2 + 885p - 87{,}025
\end{aligned}$$

Determining the domain for composite functions can be a little tricky. For example, let $p(x) = \sqrt{x - 3}$ and $q(x) = \dfrac{x + 1}{x - 3}$. It is easy to see that the domain for $p(x)$ is $x \geq 3$, while the domain for $q(x)$ is $x \neq 3$. What is the domain for the composed function $r(x) = q(p(x))$? We begin with the domain of the inner function, so $x \geq 3$. We now have to consider if there are restrictions on the outer function. In this case, we cannot just say $x \neq 3$ but must determine the value of x for which $q(x) = 3$. Solve $\sqrt{x - 3} = 3$ to determine that $x \neq 9$. Therefore, the domain for $r(x)$ is $x \geq 3$ and $x \neq 9$.

A much more interesting problem is when the two functions are $g(x) = x^2$ and $h(x) = \sqrt{x}$. Consider the two composite functions $f(x) = g(h(x))$ and $k(x) = h(g(x))$. The domain of $h(x)$ is $x \geq 0$ the domain of $g(x)$ are all values of x. Consequently, the domain for $f(x)$ is $x \geq 0$. The domain for $h(x)$ is the set of all real numbers because the output values for $g(x)$ satisfy the domain for $h(x)$. This is probably the best example of why you should determine the domain of the composed function before simplifying the expression that is the composed function. What does this mean? $\left(\sqrt{x}\right)^2$ is really just x since $x \geq 0$, and $\sqrt{x^2}$ simplifies to be $|x|$ since the domain is the set of real numbers.

EXAMPLE

▶ Given the functions $b(x) = \dfrac{x + 2}{4x - 1}$ and $c(x) = \dfrac{x - 2}{2x + 1}$, determine the domain for the function $d(x) = b(c(x))$.

▶ The domain for $c(x)$ is $x \neq \dfrac{-1}{2}$ and the domain for $b(x)$ is $x \neq \dfrac{1}{4}$. We need to determine for what value of x, if any, does $c(x) = \dfrac{x - 2}{2x + 1} = \dfrac{1}{4}$.

▶ Solve the equation:
$$4(x - 2) = 2x + 1 \Rightarrow 4x - 8 = 2x + 1 \Rightarrow 2x = 9 \Rightarrow x = \dfrac{9}{2}.$$

▶ Consequently, the domain for $d(x)$ is $x \neq \dfrac{-1}{2}, \dfrac{9}{2}$.

Inverse Functions

An inverse function is one that satisfies the equation $f(g(x)) = x$ for all values of x in its domain. In many algebraic examples, the process for determining the inverse function is to interchange the x and y in the equation and then solve for y. For example, the inverse of the function $y = f(x) = 2x + 1$ is found by solving the equation $x = 2y + 1$ for y to determine that $y = f^{-1}(x) = \dfrac{x-1}{2}$.

You will recall that there are some issues that arise that force us to put restrictions on our answers. For example, the function $f(x) = x^2$ only has an inverse if we restrict the domain to $x \geq 0$ (because the solution to the equation $x = y^2$ is $x = \pm y$ and we can only have one solution to be a function).

Another problem you experienced in Algebra II is that the inverse of an exponential function is a logarithmic function and that had to be by definition because we had no algebraic means to solve the equation $b^y = x$. We'll take a look at a few examples for inverse functions in this section and will note there are more cases to come in the following chapters of this book.

EXAMPLE

▶ Determine the inverse of the function $f(x) = \dfrac{x-1}{2x+1}$.

▶ Interchange the variables x and y: $y = f(x) = \dfrac{x-1}{2x+1} \Rightarrow x = \dfrac{y-1}{2y+1}$

▶ Multiply: $x(2y+1) = y - 1$

▶ Solve for y: $2xy + x = y - 1$

$$2xy - y = -x - 1$$

$$y(2x-1) = -x - 1$$

$$y = f^{-1}(x) = \dfrac{-x-1}{2x-1}$$

This brings us to a couple of important points. We read earlier in this chapter that the domain of the inverse function must be the same set as the range of the original function. Since the domain of the inverse of $f(x)$ is $x \neq \dfrac{1}{2}$, the range of the function $y = f(x)$ must be $y \neq \dfrac{1}{2}$.

The second point regards the existence of an inverse. As we observed when we examined the function $f(x) = x^2$, the inverse function only exists if there is a unique input value for each output value (not to be confused with the definition of a function that says there must be a unique output value for each input value). Graphically, we can determine if the graph represents a function if it passes the vertical line test. We can now say that the graph represents a relation that has an inverse function only if it passes the horizontal line test. Furthermore, a relation whose graph can pass both the vertical and horizontal line tests is called a one-to-one function (often represented as $1 - 1$).

The graph of the function $f(x) = x + \dfrac{1}{x-2}$ shows that it is not a $1 - 1$ function since it fails the horizontal line test, so no inverse exists.

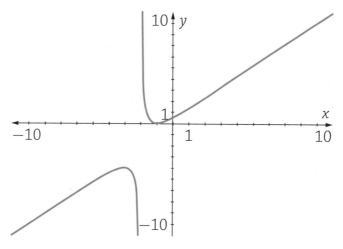

EXERCISES

EXERCISE 1-1

Given the relation $C = \{(-4, 3), (-2.5), (0, 3), (2, -9), (5, -1)\}$,

1. Determine the domain of C.

2. Determine the range of C.

3. Determine the inverse of C.

4. Is C a function? Explain.

5. Is C^{-1} a function? Explain.

Let $f(x) = -2x^2 + 50$, $g(x) = \dfrac{5x + 7}{x + 4}$, and $h(x) = \sqrt{16 - 7x}$. Use these functions to evaluate questions 6 to 14. Use these definitions to answer questions 15 to 17.

6. $f(3)$

7. $f(-3)$

8. $g(9)$

9. $g(-2)$

10. $h(1)$

11. $h(-15)$

12. $f(3) + h(-15)$

13. $\dfrac{f(3)}{h(1)}$

14. $f(-3) + 3g(9) - 2h(-15)$

15. Determine the domain of $f(x)$.

16. Determine the domain of $g(x)$.

17. Determine the domain of $h(x)$.

EXERCISE 1-2

Use the functions $m(x) = \dfrac{x-3}{x+4}$, $n(x) = \sqrt{4-2x}$, and $p(x) = 2x^2$ to answer questions 1 to 7.

1. Evaluate $m\big(n(-16)\big)$

2. Evaluate $m\big(p(2)\big)$

3. Evaluate $p\big(n(m(-3))\big)$

4. Determine the domain of $p(n(x))$.

5. Determine the domain of $n(p(x))$.

6. Determine the rule for $p(n(x))$.

7. The number of ball bearings produced at the Apex ball bearing plant t hours after business opens is given by the function $n(t) = -0.5t^2 + 4t$ and the cost for producing q ball bearings is given by $C(q) = 0.005q + 400$. Explain the meaning of the expression $C(n(3))$.

EXERCISE 1-3

Answer these function-related questions.

1. Determine the rule for $f^{-1}(x)$ if $f(x) = 5 - 7x$.

2. Determine the rule for $g^{-1}(x)$ if $g(x) = \dfrac{5+3x}{2x-7}$.

3. Determine the range of the function $g(x) = \dfrac{5+3x}{2x-7}$.

4. Determine the rule for $g^{-1}(x)$ if $h(x) = \dfrac{ax+b}{cx+d}$.

5. Determine the range of the function $h(x) = \dfrac{ax+b}{cx+d}$.

6. The graph of the function $y = f(x)$ is shown. Is the inverse of this function also a function? Explain.

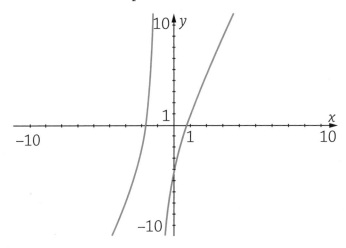

Quadratic Functions

MUST ⚡ KNOW

⚡ Both imaginary and complex numbers can be graphed on a number plane.

⚡ Complex numbers can be added and subtracted graphically.

⚡ Some optimization problems—maximum and minimum values—can be solved with quadratic functions.

side from guessing in a trial-and-error approach, you can use three ways to solve a quadratic equation: factor and use the **Zero Product Property**, use the quadratic formula, and graph the quadratic and determine the values of the x-intercepts. (If you thought to yourself that you could complete the square, well done. However, that is essentially the same as using the quadratic formula.)

Let's review some of the basic expansion/factoring relations learned in your past math classes:

- **Difference of Squares** $(a+b)(a-b) = a^2 - b^2$

- **Square Trinomials** $(a+b)^2 = a^2 + 2ab + b^2$;

  $(a-b)^2 = a^2 - 2ab + b^2$

Solving Quadratic Equations

Here's an example of solving an equation using a factoring approach.

EXAMPLE

▶ Solve the equation: $4x^3 - 100x = 0$.

▶ Remove the common factor $4x$: $4x(x^2 - 25) = 0 \Rightarrow 4x(x-5)(x+5) = 0 \Rightarrow$ $x = 0, 5, -5$.

The quadratic formula, $x = \dfrac{-b \pm \sqrt{b^2 - 4ac}}{2a}$, solves all equations of the form $ax^2 + bx + c = 0$.

Here is an example of using the quadratic formula to solve an equation.

EXAMPLE

▶ Solve the equation $-6x^2 + 25x + 16 = 0$.

▶ Using $a = -6$, $b = 25$, and $c = 16$, $x = \dfrac{-25 \pm \sqrt{25^2 - 4(-6)(16)}}{2(-6)} =$ $\dfrac{-25 \pm \sqrt{625 + 384}}{-12} = \dfrac{-25 \pm \sqrt{1009}}{-12}$.

▶ Do you see what happens if you factor −1 from the numerator and denominator? The answer becomes $x = \dfrac{25 \pm \sqrt{1009}}{12}$.

While the quadratic formula can solve all quadratic equations, you will find that there are problems with easily recognizable factors. In such cases you'll find that factoring is the faster approach. Also, the quadratic formula yields solutions in the form of radicals. It is no longer critical that the radical be reduced to its simplest form. That is, $\sqrt{27}$ is as good an answer as $3\sqrt{3}$. If a solution requires a simplified radical statement, the directions will most likely state it. More than likely, you will be required to get a decimal estimate for an irrational answer than a simplified radical form.

Graphical solutions are more likely to be used with higher-order polynomials, rational and irrational functions, and exponential equations than they are with quadratic equations. Nonetheless, graphical solutions are an option.

EXAMPLE

▶ Solve $-4.9x^2 + 18.2x + 24.1 = 0$ graphically.

▶ Enter the function $y = -4.9x^2 + 18.2x + 24.1$ into your graphing utility and sketch a graph. Determine the values of the x-intercepts using the graphing utility.

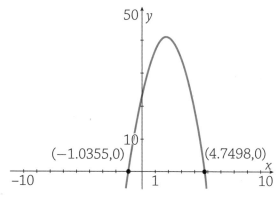

▶ The solutions are −1.036 and 4.750.

Complex Numbers

One of the "problems" that arose due to the way we talk about mathematics is that we make it sound like numbers are physical entities. Have you ever seen a 1? You have seen one finger, one cookie, or one friend, but have you ever seen a 1? No, you haven't. All numbers, like all words, are abstract entities. We have reached a common agreement about what a 1 is just as we have reached an agreement on what red is.

I raise this issue because we need to talk about numbers that are not "real." As you know, real numbers are any numbers that can be plotted on a number line. Integers are real numbers, rational numbers (the ratio of two integers in which the second integer is nonzero) are real numbers, and irrational numbers are real numbers. We come to the problem of evaluating $\sqrt{-1}$. We know that $\sqrt{4} = 2$ because $2 \times 2 = 4$. We know that any number, positive or negative, when multiplied by itself gives a positive answer (with the obvious exception of 0). So how can we have a number that, when by multiplied by itself, gives a negative answer? Why not?

We define, not compute, $\sqrt{-1}$ to equal i, and we call this number the **imaginary unit** (imaginary because it is not a *real* number). In order for this number to be accepted, there is a set of properties that it must obey. Can we perform arithmetic with these numbers, and can we represent them graphically? The answer is yes.

The arithmetic of imaginary numbers follows the same rules as do the irrational numbers, whereas $\sqrt{3} + \sqrt{3} = 2\sqrt{3}$, $\sqrt{-3} + \sqrt{-3} = \sqrt{-1}\sqrt{3} + \sqrt{-1}\sqrt{3} = i\sqrt{3} + i\sqrt{3} = 2i\sqrt{3}$. We can do the addition $\sqrt{3} + \sqrt{3} = 2\sqrt{3}$ on a number line if desired. We need to determine a way to do $\sqrt{-3} + \sqrt{-3}$ geometrically. It turns out to be a fairly easy thing to do. If we create a horizontal number line to represent the set of real numbers, a line perpendicular to this line through the origin can represent the imaginary numbers.

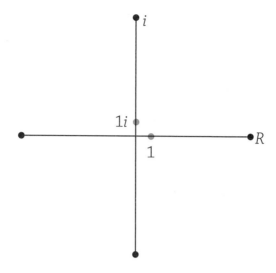

Just as we need to determine the length $\sqrt{3}$ on the real axis to perform the addition, we also need to determine the same length on the imaginary axis to perform the second addition. However, this diagram presents an interesting problem. These two intersecting lines create a number plane. What does the number with coordinates (3, 4) represent? Well, that would be 3 units to the right of the origin on the real axis and 4 units above the real axis. This combination of real and imaginary turns out to be $3 + 4i$. What kind of number is this? It's not real and it's not imaginary. Here we go again with the spoken language getting in the way of basic concepts. We have real numbers—we know what they are. We now have these imaginary numbers—we'll begrudgingly give you that because you present an interesting argument for why they should be. But $3 + 4i$? That's not a simple number. Why, then, it must a **complex number**. (I can't tell you that this was the exact conversation that led to the naming of the complex numbers, but I am willing to believe that it had to be darn close.)

Given two real numbers, a and b, and the imaginary unit i, any number of the form $a + bi$ is called an imaginary number. Notice if $b = 0$, then the number is real, but if $a = 0$, the number is imaginary. You learned in elementary school how to add two numbers a and b on a number line by using segments with their lengths and placing them end to end on the number line with one end sitting at the origin. In middle school, you learned to add signed numbers by including directions to the segments so that a positive number pointed to the right and a negative number pointed to the left. What do we do about the addition of complex numbers?

It is easy to understand that $(5 + 2i) + (2 + 3i) = 7 + 5i$; we simply add like terms. Geometrically, this becomes a very interesting, but relatively easy, problem. First, put each of these numbers in the complex number plane.

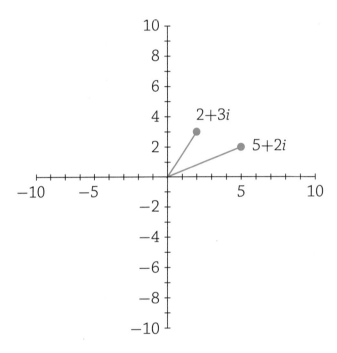

If we follow the procedure for adding on a number line, then we should put the first number in the plane, with one end on the origin, and then place the second number at the end of the first, remembering to include the orientation of the second number. Doing that, we get this picture.

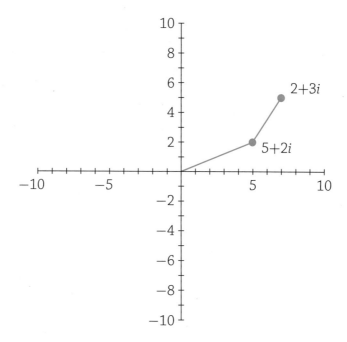

Connect the origin to the end of the second segment. As expected, the result is $7 + 5i$.

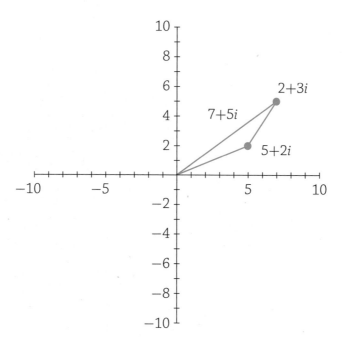

The process is called the **parallelogram method** because we take the two original segments with the angle between them and create a parallelogram. The solution to the problem is the diagonal of the parallelogram that has the origin as one endpoint.

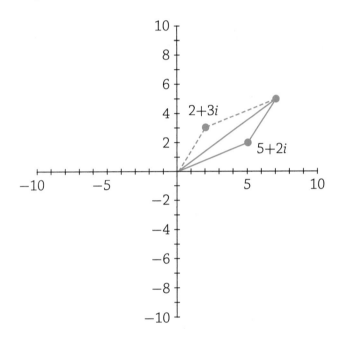

Great! We can do addition geometrically, so we've satisfied some very basic conditions. For now, let's just concentrate on the algebraic process for doing arithmetic. As was noted earlier, addition and subtraction of complex numbers are done by combining like terms.

▶ Simplify $\left(28 + 5\sqrt{-32}\right) - \left(9 - 3\sqrt{-8}\right)$.

▶ Break down the radicals and combine like terms:

$$\left(28 + 5\sqrt{-32}\right) - \left(9 - 3\sqrt{-8}\right)$$
$$\Rightarrow \left(28 + 5\sqrt{-1}\sqrt{16}\sqrt{2}\right) - \left(9 - 3\sqrt{-1}\sqrt{4}\sqrt{2}\right)$$

$$\Rightarrow \left(28 + 20i\sqrt{2}\right) - \left(9 - 6i\sqrt{2}\right).$$
$$\Rightarrow 28 + 20i\sqrt{2} - 9 + 6i\sqrt{2}$$
$$\Rightarrow 19 + 26i\sqrt{2}.$$

The rules for multiplication and division of complex numbers are exactly like those for irrational numbers.

EXAMPLE

▶ Simplify $\left(12 + 5\sqrt{-24}\right)\left(11 - 5\sqrt{-216}\right)$

▶ Break down the radicals:

$$\left(12 + 5\sqrt{-24}\right)\left(11 - 5\sqrt{-216}\right)$$
$$\Rightarrow \left(12 + 5\sqrt{-1}\sqrt{4}\sqrt{6}\right)\left(11 - 5\sqrt{-1}\sqrt{36}\sqrt{6}\right)$$
$$\Rightarrow \left(12 + 20i\sqrt{6}\right)\left(11 - 30i\sqrt{6}\right)$$

▶ Multiply the binomials:

$$\left(12 + 20i\sqrt{6}\right)\left(11 - 30i\sqrt{6}\right)$$
$$\Rightarrow 132 - 360i\sqrt{6} + 220i\sqrt{6} - 600i^2\left(\sqrt{6}\right)^2$$
$$\Rightarrow 132 - 140i\sqrt{6} + 600(6)$$
$$\Rightarrow 132 - 140i\sqrt{6} + 3600$$
$$\Rightarrow 3732 - 140i\sqrt{6}$$

Before doing an example of the division of complex numbers, recall that the numbers $a + \sqrt{b}$ and $a - \sqrt{b}$ are called **conjugates** because when they are added and multiplied, the answer does not contain radicals. The complex numbers $a + bi$ and $a - bi$ are called **complex conjugates** for a similar reason; the results will not contain imaginary numbers.

▶ Simplify $\dfrac{4 - 3i\sqrt{2}}{5 + 2i\sqrt{2}}$

▶ Begin by multiplying the numerator and denominator of the fraction by the complex conjugate of the denominator.

$$\frac{4 - 3i\sqrt{2}}{5 + 2i\sqrt{2}}\left(\frac{5 - 2i\sqrt{2}}{5 - 2i\sqrt{2}}\right) \Rightarrow \frac{20 - 8i\sqrt{2} - 15i\sqrt{2} + 6i^2(2)}{25 - \left(2i\sqrt{2}\right)^2}$$

$$\Rightarrow \frac{20 - 23i\sqrt{2} - 12}{25 + 8} \Rightarrow \frac{8 - 23i\sqrt{2}}{33}$$

▶ Complex numbers are written in the form $a + bi$, so this answer is written as $\dfrac{8}{33} - \dfrac{23i\sqrt{2}}{33}$.

One last item for this section on complex numbers: two complex numbers $a + bi$ and $c + di$ are equal if and only if $a = c$ and $b = d$.

Graphing Quadratic Functions – Parabolas

The two common forms for the equation of a parabola are $f(x) = ax^2 + bx + c$, the standard equation, and $y - k = a(x - h)^2$, the vertex form for the equation. Recall from your study of Algebra II the equation for the axis of symmetry is $x = \dfrac{-b}{2a}$ for the standard form and $x = h$ for the vertex form. In addition, the coordinates of the vertex are $\left(\dfrac{-b}{2a}, f\left(\dfrac{-b}{2a}\right)\right)$ in the standard form and (h, k) in the vertex form. Let's be real here—knowing how to determine the coordinates of the vertex in the vertex form is important. That is why it is called the vertex form. However, there is no need to know $\left(\dfrac{-b}{2a}, f\left(\dfrac{-b}{2a}\right)\right)$, but you do need to know what it means. Find the equation for the axis of symmetry and substitute that value into the function to determine the y-coordinate for the vertex.

The range of the values for the quadratic function is dependent on the location of the vertex and whether the parabola opens up or down. Let me restate that last part using the language of calculus. If the parabola is concave up (opens up), the y-coordinate of the vertex represents the minimum value of the range of the function, whereas if the parabola is concave down, the y-coordinate represents the maximum value of the range of the parabola. We can determine the concavity of the parabola by examining the sign of the leading coefficient, a. If $a > 0$, the parabola is concave up. If $a < 0$, the parabola is concave down. (If $a = 0$, the function represents a linear expression, so there is no extreme value.)

EXAMPLE

▶ Determine the range of the function $f(x) = -3x^2 + 18x + 17$.

▶ With the leading coefficient being negative, the graph of the parabola will be concave down. The equation of the axis of symmetry is $x = \dfrac{-18}{2(-3)} = 3$. The y-coordinate of the vertex is $f(3) = -3(3)^2 + 18(3) + 17 = 44$.

▶ Therefore, the range of the function is $y \leq 44$.

As you recall, finding all this information from the vertex form of the parabola is much easier.

EXAMPLE

▶ Determine the range of the function $y + 4 = 5(x - 2)^2$.

▶ The graph is concave up because the leading coefficient is positive and the coordinates of the vertex are $(2, -4)$.

▶ Therefore, the range of the function is $y \geq -4$.

Applications of Quadratic Functions

We can use the process of finding the range of the parabola to solve some very important problems. We are going to use three variations of a simple scenario to give you a feel of what the process can do.

The basis of our problems: Brendon and his sons plan to fence in a rectangular garden. The first scenario is this:

EXAMPLE

▶ If Brendon and Cameron have 80 feet of fencing, what dimensions will the garden be to maximize the area used?

▶ You might already know that the answer is 20 by 20 from your knowledge of geometry. If so, good for you! That is awesome.

▶ Let's draw a picture and get a handle on the problem. We have a rectangle with length l and width w.

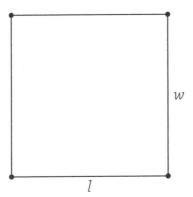

▶ We know that the perimeter of the rectangle is $2l + 2w$ and that the perimeter is the 80 feet of fencing. Therefore, $2l + 2w = 80$. We also know that the area of the rectangle is the product of the length and the width, so $A = lw$. Use the perimeter equation to isolate one of the variables: $2l + 2w = 80 \Rightarrow 2l = 80 - 2w \Rightarrow l = 40 - w$.

▶ Substitute this value for l into the area equation:

$$A = (40 - w)w = 40w - w^2 = -w^2 + 40w.$$

▶ Determine that the equation for the axis of symmetry is $w = \dfrac{-40}{2(-1)} = 20$.

This tells us that the maximum area will occur when the width is 20 feet.

▶ Use the equation for length to determine that the length must also be 20 feet.

There are two variations of this problem that we need to discuss. The first involves the use of an existing boundary.

▶ Brendon and Carson have 80 feet of fencing, but they plan to use an existing stone wall as one of the boundaries of the garden, so they will not need to fence that side. What are the dimensions of the garden with maximum area?

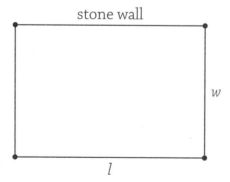

▶ While the area of the garden is still computed with the formula $A = lw$, the 80 feet of fencing is used for only three of the sides of the rectangle. The equation representing the fencing used is $2w + l = 80$ so that $l = 80 - 2w$.

Substituting for l in the area equation, we get $A = (80 - 2w)w = -2w^2 + 80w$. The axis of symmetry has the value $w = \dfrac{-80}{2(-2)} = 20$. The length of the garden is $l = 80 - 2(20) = 40$ feet.

The dimension of the garden with maximum area is 20 by 40 feet.

The last variation of this problem has a slight twist.

EXAMPLE

Brendon and Colin plan to build a rectangular garden using a (straight) creek bed as one side of the garden. The edge of the garden along the creek bed requires a stronger fence than do the other three sides. When Brendon and Colin go to the store to buy fencing, they find that the fencing along the creek bed costs \$4 per foot, while the fencing for the remaining sides costs \$3 per foot. If they have \$80 to buy the fencing, determine the dimensions of the garden with maximum area that can be enclosed for \$80.

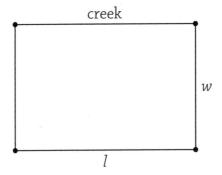

Once again, the area of the garden is found by the formula $A = lw$. However, rather than deal with the amount of fence that forms the perimeter, we deal with the cost of the fencing forming the perimeter. The side along the creek will cost $5l$ dollars, while the other three sides will cost a total of $3(l + 2w)$. Thus, the equation for the amount of fencing needed can be written as $5l + 3(l + 2w) = 80$. Simplified, this becomes $8l + 6w = 80$.

▶ As before, solve for l to get $l = \dfrac{80 - 6w}{8} = 10 - \dfrac{3}{4}w$.

▶ Substitute this into the equation for area:

$A = \left(10 - \dfrac{3}{4}w\right)w = \dfrac{-3}{4}w^2 + 10w$. The axis of symmetry for the graph of this function has equation $w = \dfrac{-10}{2\left(\dfrac{-3}{4}\right)} = \dfrac{-10}{\dfrac{-3}{2}} = \dfrac{20}{3}$. The length of the garden will be $l = 10 - \dfrac{3}{4}\left(\dfrac{20}{3}\right) = 5$.

▶ The dimensions of the garden with maximum area are $6\dfrac{2}{3}$ by 5 feet.

EXERCISES

EXERCISE 2-1

Solve each of the following equations by factoring.

1. $4x^2 - 15x - 25 = 0$

2. $36x^2 - 121 = 0$

3. $8x^2 + 22x - 21 = 0$

4. $48x^3 - 243x = 0$

5. $10x^2 - 39x - 27 = 0$

Solve each of the following equations using the quadratic formula. (Express irrational results in both radical form and decimal form, rounded to the nearest thousandth.)

6. $10x^2 - 39x - 26 = 0$

7. $8x^2 + 22x - 19 = 0$

8. $5x^2 - 19x + 13 = 0$

9. $-12x^2 + 11x - 2 = 0$

10. $\dfrac{1}{3}x^2 + \dfrac{3}{7}x - \dfrac{7}{9} = 0$

Solve each of the following equations graphically.

11. $\dfrac{1}{3}x^2 + \dfrac{3}{7}x - \dfrac{7}{9} = 0$

12. $4x^2 - 9x - 12 = 0$

13. $-3x^2 + 12x - 11 = 0$

14. $5x^2 - 20x + 17 = 0$

15. $-6x^2 + 24x - 25 = 0$

EXERCISE 2-2

Compute each of the following.

1. $\left(18 + 19\sqrt{-18}\right) + \left(15 - 10\sqrt{-50}\right)$

2. $\left(81 + 91\sqrt{-48}\right) - \left(51 - 40\sqrt{-75}\right)$

3. $\left(8 + 11\sqrt{-18}\right)\left(12 - 4\sqrt{-50}\right)$

4. $\left(8 + 11\sqrt{-24}\right)^2$

5. $\left(17 + 2\sqrt{-13}\right)\left(17 - 2\sqrt{-13}\right)$

6. $\dfrac{17 + 2\sqrt{-3}}{8 - 4\sqrt{-3}}$

Solve each equation for x.

7. $x + 13 + 2i\sqrt{3} = 11 - 3i\sqrt{12}$

8. $(4 + 3i)x = 5 - 2i$

9. $\dfrac{x}{2 + 3i\sqrt{2}} = 5 - 4i\sqrt{8}$

10. $\dfrac{2 + 3i\sqrt{2}}{x} = 5 - 4i\sqrt{8}$

EXERCISE 2-3

For questions 1 to 5, determine the range of the given function.

1. $f(x) = -2x^2 - 12x + 19$

2. $y - 10 = \dfrac{2}{3}(x + 6)^2$

3. $g(x) = \dfrac{3}{5}x^2 - 12x + 19$

4. $y = \dfrac{\sqrt{3}}{2}(x - 5)^2 + 7$

5. $g(x) = \dfrac{-3}{7}x^2 + 21x + 109$

6. The profit earned by BeeMax Honey is given by the function $P(g) = -0.02g^2 + 6g - 25$, where g represents the number of gallons of honey produced in a day. Determine the maximum profit that BeeMax can earn in one day.

7. When a ball is thrown vertically into the air with an initial speed of 96 feet per second from a point 6 feet above the ground, the height of the ball can be determined by the function $h(t) = -16t^2 + 96t + 6$, where t represents the number of seconds since the ball has been thrown. Determine the maximum height of the ball.

8. Brendon and Cameron have 180 feet of fencing to enclose a rectangular garden. What are the dimensions of the garden with maximum area?

9. Brendon and Carson have 180 feet of fencing, but they plan to use an existing stone wall as one of the boundaries of the garden so they will not need to fence that side. What are the dimensions of the garden with maximum area?

10. Brendon and Colin plan to build a rectangular garden using a (straight) creek bed as one side of the garden. The edge of the garden along the creek bed requires a stronger fence than do the other three sides. When Brendon and Colin go to the store to buy fencing, they find that the fencing along the creek bed costs $5 per foot, while the fencing for the remaining sides costs $2 per foot. If they have $200 to buy the fencing, determine the dimensions of the garden with maximum area that can be enclosed for $200.

Flashcard App

Polynomial Functions

MUST KNOW

⚡ The Remainder and Factor theorems are important tools in determining the factors of a polynomial.

⚡ It is often easier to identify the factors of a polynomial by examining the related graph.

⚡ The Fundamental Theorem of Algebra tells us that for any polynomial $f(x)$ of degree n, there are n roots from the set of complex numbers that solve the equation $f(x) = 0$.

⚡ Signs analysis is a key method for solving polynomial inequalities.

n an age where computing devices run rampant, there is still some reluctance among many educators to employ calculators with **Computer Algebra Systems (CAS)** as part of their curriculum. Much of the material in this chapter can be easily done with the assistance of such a device. There are a few examples in the chapter that are similar to what might appear as part of a calculus problem that are worth looking at should you be one of the lucky students who use a CAS device.

Factoring Polynomials

Factoring polynomials with a degree greater than 3 (the degree of a polynomial is the largest exponent in the expression) is usually very difficult to do unless the expression fits some special pattern. We'll take a look at some of the patterns in this section and then move on to other techniques for working with polynomials.

We know that the difference of squares will factor ($x^2 - y^2 = (x + y) \cdot (x - y)$), as does the difference of cubes, ($x^3 - y^3 = (x - y)(x^2 + xy + y^2)$). It is a fact that the difference of nth powers will always factor ($x^n - y^n = (x - y) \cdot (x^{n-1} + x^{n-2}y + x^{n-3}y^2 + \ldots + xy^{n-2} + y^{n-1})$).

EXAMPLE

▶ Factor $x^7 - y^7$.

▶ $x^7 - y^7 = (x - y)(x^6 + x^5y + x^4y^2 + x^3y^3 + x^2y^4 + xy^5 + y^6)$

We know that the sum of squares will not factor over the real numbers but that the sum of cubes will ($x^3 + y^3 = (x + y)(x^2 - xy + y^2)$). It is a fact that the sum of nth powers will factor provided that n is an odd number ($x^n + y^n = (x + y)(x^{n-1} - x^{n-2}y - x^{n-3}y^2 - \ldots - xy^{n-2} + y^{n-1})$).

EXAMPLE

▶ Factor $x^5 + y^5$.

▶ $x^5 + y^5 = (x + y)(x^4 - x^3y - x^2y^2 - xy^3 + y^4)$

It is true the expression $x^6 + y^6$ factors, but this is only because we can rewrite the expression to be the sum of two cubes, $(x^6 + y^6 = (x^2)^3 + (y^2)^3)$.

It is true that we can be asked to perform some mental gymnastics to prove that we can remember the factoring patterns for the difference or sum of higher-order polynomials $(x^7 - 32x^2 = x^2(x^5 - 2^5) = x^2(x - 2)(x^4 + 2x^3 + 4x^2 + 8x + 16))$. It is not all that often that we will need to remember this, as the material that follows in this chapter will cover these special cases.

Remainder and Factor Theorems

Let me take you back to elementary school when you first learned how to do division. You learned that 42 divided by 7 was equal to 6 because 7 times 6 was 42. (The modern nomenclature is that 6, 7, and 42 form a factor family.) About the same time, you learned the vocabulary word **factor**: 6 and 7 are factors of 42 (though not the only factors). You then learned that there are such things as remainders and that when the remainder of a division problem is 0, you call the divisor a factor of the dividend.

We can extend that material to polynomial functions. If $f(x)$ is a polynomial function, then when $f(x)$ is divided by the linear expression $x - a$, there will be a quotient $Q(x)$ and a remainder R. That is,

$$\frac{f(x)}{x - a} = Q(x) + \frac{R}{x - a}$$

Multiply both sides of the equation by $x - a$:

$$f(x) = Q(x)(x - a) + R$$

If we evaluate the function $f(x)$ with $x = a$, we get $f(a) = R$. That is, the value of the function at a particular value gives the remainder when the function is divided by the linear expression involving that value. This is called the **Remainder Theorem**. Of special interest is when the remainder is equal to 0. This indicates that the linear expression is a factor of the polynomial. Naturally, this is called the **Factor Theorem**.

EXAMPLE

▶ Determine if $x - 4$ is a factor of the polynomial $3x^4 + 2x^3 - 39x^2 - 62x - 24$.

▶ Define $f(x)$ as $3x^4 - 17x^3 + 28x^2 - 26x - 24$ and determine that $f(4) = 0$, so $x - 4$ is a factor of $3x^4 + 2x^3 - 39x^2 - 62x - 24$.

Synthetic Division

We saw in the previous section that $x - 4$ is a factor of $3x^4 + 2x^3 - 39x^2 - 62x - 24$. The question that should come to mind is "What is the other factor?" We could go through the grueling process of long division with $3x^4 + 2x^3 - 39x^2 - 62x - 24$ and $x - 4$, but that is more work than I'd like to do. If the course you are taking allows for graphing calculators, you can have your device do the work for you.

$$f(x) \qquad\qquad 3x^4 + 2x^3 - 39x^2 - 62x - 24$$

$$\frac{f(x)}{x - 4} \qquad\qquad 3x^3 + 14x^2 + 17x + 6$$

Now that we have calculators with CAS, we would simply type the command factor $(3x^4 + 2x^3 - 39x^2 - 62x - 24)$ and the device will display the result. The material that follows involving the factoring is for those whose graphing calculators do not use a CAS and also contains some important information regarding the process of finding the roots of a polynomial.

If you do not have the use of a graphing calculator (or some other electronic device), there is a technique called synthetic division that will enable you to do the division much more efficiently than using long division. To begin, we solve the equation $x - 4 = 0$ to determine that $x = 4$. We'll use this value and the coefficients of $f(x)$ to solve the problem.

$$\underline{4|}\ \ 3\ \ \ 2\ \ -39\ \ -62\ \ -24$$

We bring down the leading coefficient, multiply this number by the divisor, and write the product under the next coefficient.

$$
\begin{array}{r}
\underline{4|}\ \ 3\ \ \ 2\ \ -39\ \ -62\ \ -24 \\
12 \\
\hline
3
\end{array}
$$

Add the product to the coefficient, writing the sum under the line.

$$
\begin{array}{r}
\underline{4|}\ \ 3\ \ \ 2\ \ -39\ \ -62\ \ -24 \\
12 \\
\hline
3\ \ 14
\end{array}
$$

We will repeat the multiply, record the product under the next coefficient, and proceed until we have used all the coefficients.

$$4 \underline{|\ 3\ \ 2\ -39\ -62\ -24}$$
$$12\ \ \ \ 56\ \ \ \ 68\ \ \ \ 24$$
$$\overline{\ 3\ \ 14\ \ \ \ 17\ \ \ \ \ 6\ \ \ \ \ \ 0}$$

Compare the last line of the division with the coefficients of the quotient, $3x^3 + 14x^2 + 17x + 6$, found with our calculator. The 0 is the remainder, while the other values are the coefficients of the quotient.

There are three items we need to discuss. The first is easy. What if the divisor is $x + a$? Simple. Solve the equation $x + a = 0$ and go about your process.

The second item is a little more challenging. What if there is a missing term in the polynomial? For example, what if the polynomial is $3x^3 - 14x^2 + 6$? We use a 0 as the coefficient of the missing term. That is, the first line in the synthetic division will be 3 −14 0 6.

The third item might be the trickiest of the bunch. What if the divisor is of the form $bx - a = 0$? We still solve for x and use that value in our synthetic division. However, we have the extra step of dividing the coefficients of the quotient by b to determine the final answer.

EXAMPLE

▶ Divide $4x^4 - 73x^2 + 144$ by $2x + 3$.

▶ Solve $2x + 3 = 0$ to get $x = \dfrac{-3}{2}$.

▶ Set up the synthetic division using 0 as a coefficient for x^3 and for x.

$$\dfrac{-3}{2} \underline{\Big|\ 4\ \ 0\ -73\ \ 0\ \ 144}$$
$$\overline{}$$

▶ Perform the synthetic division.

$$\dfrac{-3}{2}\Big|\ \ \begin{array}{rrrrr} 4 & 0 & -73 & 0 & 144 \\ & -6 & 9 & 96 & -144 \\ \hline 4 & -6 & -64 & 96 & 0 \end{array}$$

▶ We have $\dfrac{4x^4 - 73x^2 + 144}{2\left(x + \dfrac{3}{2}\right)} = \dfrac{4x^3 - 6x^2 - 64x + 96}{2}$. Therefore, the

quotient when $\dfrac{4x^4 - 73x^2 + 144}{2x + 3}$ is $2x^3 - 3x^2 - 32x + 48$.

Graphing Polynomial Functions

There are a few things we need to talk about regarding the graphs of polynomial functions that are not necessarily obvious from a graphing calculator. These include the concepts of even and odd functions, multiplicities, and end behavior.

Polynomials such as $f(x) = x^2$, $f(x) = x^4$, $f(x) = x^6$, and $f(x) = x^4 - 5x^2 + 3$ have the property that they are symmetric to the y-axis. The reason is that all the exponents in the polynomial are even numbers (constants at the end of the polynomial such as $f(x) = x^4 - 5x^2 + 3$ can be thought of as $f(x) = x^4 - 5x^2 + 3x^0$). For that reason, these functions are called even functions. An advantage of these functions is that if you know the value $f(a) = b$, you also know the value of $f(-a) = b$.

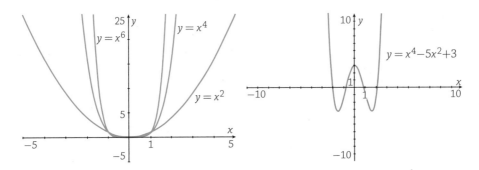

Polynomials that have only odd exponents are called odd functions. These polynomials will be symmetric to the origin. Some examples are $f(x) = x$, $f(x) = x^3$, $f(x) = 4x^5$, and $f(x) = x^5 - 8x^3 + 3x$. With odd functions, knowing that the value of $f(a) = b$ also tells you that $f(-a) = -b$.

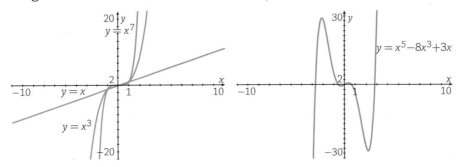

Polynomials that have a combination of even and odd exponents are designated under neither category and, as you might well imagine, are the more commonly seen polynomial functions.

Observe the difference between the graphs of $f(x) = x^2$, $f(x) = x^4$, and $f(x) = x^6$. The larger the exponent, the "flatter" the graph appears around the origin. The same can be seen with the graphs of $f(x) = x^3$ and $f(x) = 4x^5$. You know that when you solve the equation $(x - 2)^2 = 0$ we write the solution as $x = 2$, but the real solution is $x = 2$ or $x = 2$. There are two solutions to the problem, but we only write the solution once. In a similar way, the solution to the equation $(x - 2)^3 = 0$ is $x = 2$ or $x = 2$ or $x = 2$. These solutions are called **multiplicities**. What impact do multiplicities have on the graph of a function? They "flatten" the function around the point that is the multiplicity. Look at the graphs of $y = (x - 2)^4$ and $y = (x - 2)^5$.

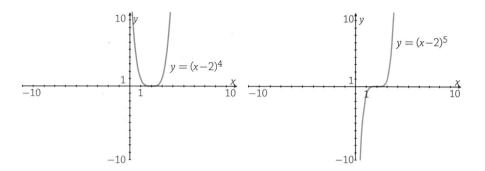

The even function "bounces" off the x-axis (that is, it is tangent to the x-axis), while the odd function continues to ride through the x-axis. This is very helpful to know when you are asked to graph polynomials and also when you use the graph to factor a polynomial. (More on the factoring aspect shortly.)

The third item mentioned is that of end behavior. What we mean by this is what direction does the graph go as x grows very large in magnitude in both the negative and positive directions? Or, in mathematical terms, in what direction does the graph move as x approaches negative infinity and in what direction does the graph move as x approaches positive infinity?

Because polynomials can have as many terms in it as we'd like, this might seem like a rather daunting issue. It turns out, thankfully, that this is not the case. The only term we need consider is the term with the highest degree and the coefficient that goes with it. If the degree of the polynomial is an even number, then the graph will approach infinity as x gets infinitely large in magnitude *provided* the coefficient of the largest term is positive. If the coefficient is negative, the graph will move toward negative infinity. If the degree of the polynomial is odd and the leading coefficient is positive, the graph will move toward negative infinity as x goes to negative infinity and will go to positive infinity as x goes to positive infinity. The behaviors will reverse if the coefficient is negative. The end behaviors are summarized in the following table.

Even function, and leading coefficient is positive.	Even function, and leading coefficient is negative.
$x \to -\infty \Rightarrow y \to \infty$ $x \to \infty \Rightarrow y \to \infty$ 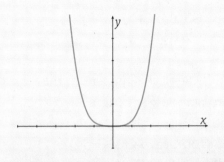	$x \to -\infty \Rightarrow y \to -\infty$ $x \to \infty \Rightarrow y \to -\infty$ 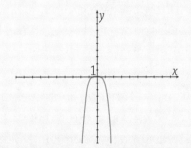
Odd function, and leading coefficient is positive.	Odd function, and leading coefficient is negative.
$x \to -\infty \Rightarrow y \to -\infty$ $x \to \infty \Rightarrow y \to \infty$ 	$x \to -\infty \Rightarrow y \to \infty$ $x \to \infty \Rightarrow y \to -\infty$ 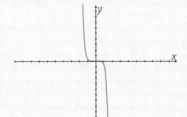

EXAMPLE

▶ Describe the end behavior of the function $f(x) = -4x^5 + 18x^4 - 20x^3 + 1000x + 15$.

▶ The large coefficients for the three middle terms might cause you to pause for a moment, but remember, it is only the term of highest degree, in this case $-4x^5$, that will dictate the end behavior.

▶ The degree of the polynomial is odd and the coefficient is negative, so $x \to -\infty \Rightarrow y \to \infty$ and $x \to \infty \Rightarrow y \to -\infty$.

Using Graphs to Factor Polynomials

We can use the relationship among roots, factors, and x-intercepts to factor polynomials. An important skill to know is how to use your graphing calculator to find the zeroes of a function. (As many different calculators on the market have this ability, it is not reasonable to review all the processes here. The biggest issue will be to find the menu on your device that contains the Zero command.)

> **BTW**
>
> *When you want to find the factors of a polynomial, it is always a good idea to add the coefficients of the polynomial together. If they sum to zero, then you know from the Factor theorem that $x – 1$ is a factor of the polynomial.*

Let's work through the problem of factoring $x^4 - 23x^2 - 18x + 40$. The sum of the coefficients is zero, so we know that $x - 1$ is a factor of the polynomial. Graph the polynomial on your device. Keep in mind that this is often an issue for new users of graphing devices, because using incorrect window dimensions can distort how the graph looks. For the purposes of finding the zeroes, it is more important to be able to identify an interval for the x values that show the zeroes than to have the y values showing you the "big" picture. Look at the difference between these two graphs.

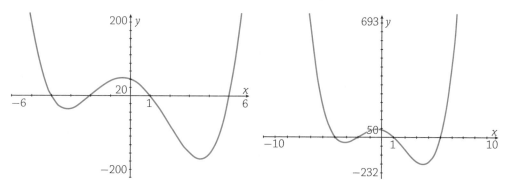

The graph on the left gives more room between the zeroes than does the graph on the right. The graph on the right certainly illustrates the end behavior of the function, but that is not a concern at this point.

Let's continue. We can see that there are four x-intercepts and we need to determine each of them by repeating the same set of steps. The computing device is going to ask us to create an interval containing the zero. I find it easiest to look at the graph and pick an integer on either side of the intercept—with the provision that there is only one intercept in the interval: the leftmost intercept.

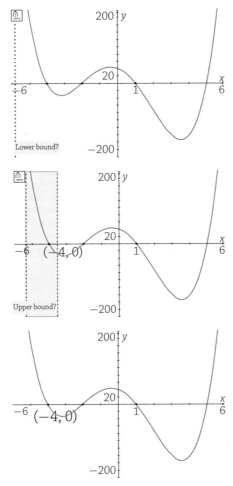

We can see that −4 is an x-intercept, so x = −4 must be a root and x + 4 must be a factor of the polynomial.

Repeat this process to find that the other factors are x + 2, x − 1, and x − 5. Therefore, $x^4 - 23x^2 - 18x + 40 = (x + 4)(x + 2)(x - 1)(x - 5)$.

EXAMPLE

▶ Determine the factors of $14x^5 + 125x^4 + 336x^3 + 324x^2 - 108x - 81$.

▶ I am comfortable writing that this is about as hard a problem that you might encounter. I hope that you find the process easy to do.

▶ Get a graph of the problem.

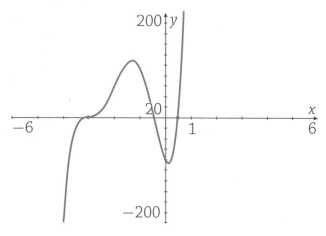

▶ It is true that we cannot see the turning point at the top of the graph, but that is not a critical item when we are just trying to determine the values of the x-intercepts. Notice that the leftmost intercept flattens out as it crosses the axis. This represents a multiplicity of three. How do we know it is three and not some other odd number? The polynomial itself is of degree five, and we clearly have single roots for the other two intercepts. That leaves three roots to find, and they are all the same. Let's get to it.

▶ Identify the three intercepts:

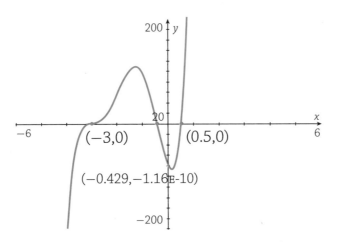

▶ The first intercept is at $x = -3$, so $x + 3$ must be a factor with multiplicity 3. The next intercept is $x = 0.5 = \dfrac{1}{2}$, so we determine that $2x = 1$ and $2x - 1$ must be a factor.

▶ The last intercept is interesting. First, we need to understand that the y-coordinate of the intercept is really -1.66×10^{-10} and that is really zero in the computer's mind (there is a lot of rounding and extra digits being carried along, and this device has a limited capacity for carrying them, so we sometimes get these little surprises).

▶ How do we address the value -0.429? It is not a value that one normally comes across, and more than likely it is a truncated version of the full decimal. Fortunately, your calculator stores the results of the intercepts in a memory location. You'll need to read your manual or use a search engine to determine how your device works. Once you have found the memory location, you can use the decimal to fraction command to determine that the last intercept is $\dfrac{-3}{7}$ and that the last factor is $7x + 3$.

▶ We can now determine that $14x^5 + 125x^4 + 336x^3 + 324x^2 - 108x - 81 = (x + 3)^3 (3x + 7) (2x - 1)$.

Let's do a similar problem, only in reverse.

EXAMPLE

Determine the function $f(x)$, whose graph is shown.

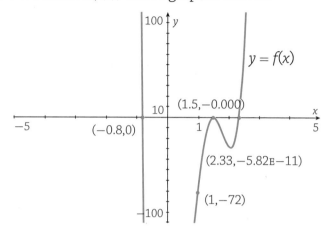

▶ The x-intercepts are -0.8, 1.5, and 2.33. (Do you see how -0.000 is another example of the round-off errors that many calculating devices incur?) The intercept at 1.5 appears to be of multiplicity 2, as the graph is tangent to the x-axis and not as flat as a graph of multiplicity 4 would be.

▶ The first intercept at $x = -0.8$ becomes the factor $4x + 5$ and the intercept at $x = 1.5$ becomes the repeated factor $2x - 3$. As we did in the last example, use the decimal to fraction command to turn the last intercept into the factor $3x - 7$.

▶ So, now we know that the function in question must be of the form $f(x) = a(5x + 4)(2x - 3)^2(3x - 7)$. There are an infinite number of functions that have these intercepts, including the multiplicity of 2 at $x = 1.5$.

▶ I am sure you saw there was an extra point identified on the graph. This is the key for determining the exact function in question. Use $f(1) = -72$ to get

$$f(1) = a(5(1) + 4)(2(1) - 3)^2(3(1) - 7) = -72$$

$$a(9)(-1)^2(-4) = -72 \Rightarrow a = 2$$

▶ Therefore, $f(x) = 2(5x + 4)(2x - 3)^2(3x - 7)$.

Fundamental Theorem of Algebra

There is a good chance that you have heard the name Carl Friedrich Gauss (1777–1855) by this point in your education. Among many things, he is credited with determining the formula for arithmetic series. Look him up on a search engine sometime. His is an interesting story. A second aspect of his work was his doctoral thesis (written while still a teen) about the roots of polynomial equations. It is known as the **Fundamental Theorem of Algebra**, and it states that a polynomial equation of degree n with real coefficients will have exactly n roots (including multiplicities) from the set of complex numbers.

We can use all the material we've discussed in this chapter to help us completely solve polynomial equations.

EXAMPLE

▶ Solve $x^3 - 1 = 0$.

▶ At first thought, one might want to say that 1 is a triple root, but that would be incorrect. One is a triple root in the equation $(x - 1)^3 = 0$. We can factor $x^3 - 1$ and get $(x - 1)(x^2 + x + 1)$.

▶ We can solve the equation $(x - 1)(x^2 + x + 1) = 0$ with the quadratic formula and determine the solution is $x = 1, \dfrac{1}{2} \pm \dfrac{\sqrt{3}}{2}i$.

This next example represents a more complicated problem.

EXAMPLE

▶ Solve $4x^5 + 16x^4 + x^3 - 105x^2 - 325x - 375 = 0$.

▶ A quick computation tells us that $x - 1$ is not a factor. Let's take a look at the graph of $y = 4x^5 + 16x^4 + x^3 - 105x^2 - 325x - 375$.

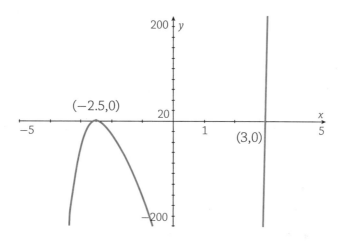

▶ The equation has a double root at $x = -2.5$ and a single root at $x = 3$. This is a fifth-degree equation, so we can use synthetic division to reduce the polynomial.

$$
\begin{array}{r|rrrrrr}
\dfrac{-5}{2} & 4 & 16 & 1 & -105 & -325 & -375 \\
 & & -10 & 15 & 35 & 175 & -375 \\
\hline
 & 4 & 6 & -14 & -70 & -150 & 0
\end{array}
$$

▶ Divide the coefficients by 2 to make the last line: 2 3 −7 −35 −75

▶ Perform the synthetic division a second time with −2.5.

$$
\begin{array}{r|rrrrr}
\dfrac{-5}{2} & 2 & 3 & -7 & -35 & -75 \\
 & & -5 & 5 & 5 & 75 \\
\hline
 & 2 & -2 & -2 & -30 & 0
\end{array}
$$

▶ Divide the coefficients by 2 to make the last line: 1 −1 −1 −15

▶ Perform the synthetic division with $x = 3$.

$$\begin{array}{r|rrrr} 3 & 1 & -1 & -1 & -15 \\ & & 3 & 6 & 15 \\ \hline & 1 & 2 & 5 & 0 \end{array}$$

▶ We now know that $4x^5 + 16x^4 + x^3 - 105x^2 - 325x - 375 = (2x + 5)^2 (x - 3)(x^2 + 2x + 5)$.

Solve $x^2 + 2x + 5 = 0$ using the quadratic formula to get $x = -1 \pm 2i$.

▶ The solution to the equation $4x^5 + 16x^4 + x^3 - 105x^2 - 325x - 375 = 0$ is $x = \dfrac{-5}{2}, 3, 1 \pm 2i$.

BTW

Devices with CAS can do this problem in one step.

cSolve $(4x^5 + 16x^4 + x^3 - 105x^2 - 325x - 375 = 0, x)$

$x = -1 + 2i$ or $x = -1 - 2i$ or $x = \dfrac{-5}{2}$ or $x = 3$

Be warned. Without a CAS device available to us, we still do not have the ability to solve many polynomial equations. For example, to solve $x^5 - 1 = 0$, we would factor the polynomial to be $x^5 - 1 = (x - 1)(x^4 + x^3 + x^2 + x + 1)$. While $x = 1$ is clearly a solution to the equation, the solution to the fourth-degree polynomial is not obvious. (FYI, the complete solution, thanks to my CAS device, is $\dfrac{-1 - \sqrt{5}}{4} \pm \dfrac{\sqrt{10 - 2\sqrt{5}}}{4}i, \dfrac{1 + \sqrt{5}}{4} \pm \dfrac{\sqrt{10 - \sqrt{5}}}{4}, 1$.)

Polynomial Inequalities

Most of this chapter has been dedicated to showing how to solve polynomial equations. We will discuss the solution of polynomial inequalities. You may have learned about a technique called **signs analysis** in Algebra II, particularly when studying quadratic inequalities. If that is so, you will find this section familiar. If you are new to this technique, we use a basic property of numbers called the trichotomy principle: a number is negative, zero, or positive and can only be one of these at any one time.

We now know how to solve a multitude of polynomial equations. We will use this knowledge to determine the sign of the polynomial in the intervals formed by these values.

EXAMPLE

▶ Solve $(2x + 5)(3x - 4)^2(5x - 12) < 0$.

▶ We know that $(2x + 5)(3x - 4)^2(5x - 12) = 0$ when $x = \dfrac{-5}{2}, \dfrac{3}{4}$, and $\dfrac{12}{5}$.

▶ For every other value of x, this polynomial will be negative or positive. We establish a number line and note where the polynomial is equal to zero.

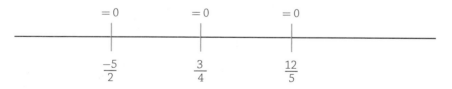

▶ For each of the intervals formed on the number line by these values, we will select a value in the interval and determine the sign for each of the factors. It is important to note that we are not interested in the magnitude of the factor, only the sign. For instance, in the interval to the far left, we can choose $x = -5$. The signs of the factors will be $(-)$ $(+)(-)$ and the resulting product will be positive.

▶ Therefore, we know that for all values of $x < \dfrac{-5}{2}$, the polynomial will have positive values. Now is a good time to note that there is a root with multiplicity 2 at $x = \dfrac{3}{4}$ so that the factor $(3x - 4)^2$ will be positive in each of the intervals on the number line.

▶ Repeat the process for the other intervals:

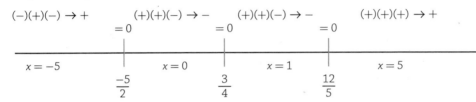

$(-)(+)(-) \to +$ $(+)(+)(-) \to -$ $(+)(+)(-) \to -$ $(+)(+)(+) \to +$

     $= 0$ $= 0$ $= 0$

$x = -5$ $\dfrac{-5}{2}$ $x = 0$ $\dfrac{3}{4}$ $x = 1$ $\dfrac{12}{5}$ $x = 5$

▶ The solution to the inequality $(2x + 5)(3x - 4)^2(5x - 12) < 0$ is $\dfrac{-5}{2} < x < \dfrac{3}{4}$ or $\dfrac{3}{4} < x < \dfrac{12}{5}$.

Let's look at a second problem.

▶ Solve $(x + 4)(x - 2)^3(x - 5)(x^2 + 9) \geq 0$.

▶ The roots are $x = -4, 2, 5$. Two is a root of multiplicity 3, so the sign of this factor will be whatever the sign of $x - 2$ is. The irreducible quadratic factor $x^2 + 9$ will always be positive. The signs analysis for this problem will look like this:

$(-)(-)(-)(+) \to -$ $(+)(-)(-)(+) \to +$ $(+)(+)(-)(+) \to -$ $(+)(+)(+)(+) \to +$

     $= 0$ $= 0$ $= 0$

$x = -5$ -4 $x = 0$ 2 $x = 3$ 5 $x = 6$

▶ The solution to the inequality $(x + 4)(x - 2)^3(x - 5)(x^2 + 9) \geq 0$ is $-4 \leq x \leq 2$ or $x \geq 5$.

Rational Root Theorem and Descartes Rule of Signs

It can be argued that up to this point the work we've done in this chapter can be done without electronic assistance (provided, of course, that we make the expressions we used easily factored). We will take a look at two statements that had great use in the days before graphing calculators.

The first of these statements is the **Rational Root Theorem**. The theorem tells us that if the polynomial $a_n x^n + a_{n-1} x^{n-1} + \ldots + a_2 x^2 + a_1 x + a_0$ has any rational roots, then those roots will be of the form $\dfrac{p}{q}$ where p is a factor of a_0 and q is a factor of a_n.

EXAMPLE

▶ Determine the set of possible rational roots for the equation $2x^4 - 15x^3 + 16x^2 + 93x - 180 = 0$.

▶ The factors of 2 are 1 and 2. The factors of 180 are 1, 180, 2, 90, 3, 60, 4, 45, 5, 36, 6, 30, 9, 20, 10, 18, and 12, 15. (Did you notice that I started with 1 and its factor, 2 and its factor, etc., until I got to the end of the list?)

▶ Let's rewrite the factors of 180 from smallest to largest: 1, 2, 3, 4, 5, 6, 9, 10, 12, 15, 18, 20, 30, 36, 45, 60, 90, 180.

▶ We need to list all the possible combinations of the ratios for the factors of 180 and 2. We also need to remember the issue of signs. We can repeat the factors of 180 when using 1 from the factor of 2, and when using the 2 from the factor of 2, we do not need to repeat fractions that reduce to a previously listed value.

$$\frac{p}{q} = \pm\left\{1,2,3,4,5,6,9,10,12,15,18,20,30,36,45,60,90,180,\frac{1}{2},\frac{3}{2},\frac{5}{2},\frac{9}{2},\frac{15}{2},\frac{45}{2}\right\}$$

Now that we have the possible rational roots, we need to find them. But before we do that, we'll take a look at **Descartes Rule of Signs**. René Descartes, the mathematician/philosopher responsible for the coordinate plane, showed that the number of positive real zeroes in a polynomial function $f(x)$ is the same as or less than by an even number as the number of changes in the sign of the coefficients and that the number of negative real zeroes in a polynomial function $f(x)$ is the same as or less than by an even numbers as the number of changes in the sign of the coefficients of $f(-x)$.

The reason for the phrase "or less than by an even number" is due to the fact that complex roots always come in conjugate pairs. Make sure you notice that the statement talks about real zeroes and not just rational zeroes. (What happens if there is a term missing in the polynomial? We treat the missing term as having a coefficient of zero, and consequently there is no change in sign.)

EXAMPLE

▶ Apply Descartes Rule of Signs to the function
$f(x) = 2x^4 - 15x^3 + 16x^2 + 93x - 180$.

▶ There are three sign changes in the coefficients of the functions, telling us that there are 3 or 1 real zeroes to the function. Given
$f(-x) = 2(-x)^4 - 15(-x)^3 + 16(-x)^2 + 93(-x) - 180 = 2x^4 + 15x^3 + 16x^2 - 93x - 180$,

you can see that there is only one sign change, so we are guaranteed that there is 1 negative real zero to the function.

We now get to play detective and go about finding whatever roots to the polynomial that we can.

▶ We'll start with $x = 1$ and $x = -1$ because all we need to do is add the coefficients: $f(1) = -84$ and $f(-1) = -240$.

▶ That wasn't much help (but it was quick). Evaluate the polynomial at $x = 2$ to get the value -18 (that's better), and, even better still, we can determine that $x = 3$ is a zero! One down, three to go. Our next step is to reduce the polynomial using synthetic division. Why? It's easier to work with smaller numbers.

$$\underline{3 |}\ \ \begin{array}{rrrrr} 2 & -15 & 16 & 93 & -180 \\ & 6 & -27 & -33 & 180 \\ \hline 2 & -9 & -11 & 60 & 0 \end{array}$$

▶ We now have
$$2x^4 - 15x^3 + 16x^2 + 93x - 180 = (x - 3)(2x^3 - 9x^2 - 11x + 60).$$

▶ Let's try $x = 3$ again:
$2(3)^3 - 9(3)^2 - 11(3) + 60 = 54 - 81 - 33 + 60 = 0$. Ah, $x = 3$ is a double root. Reduce the polynomial again.

$$\underline{3 |}\ \ \begin{array}{rrrr} 2 & -9 & -11 & 60 \\ & 6 & -9 & -60 \\ \hline 2 & -3 & -20 & 0 \end{array}$$

▶ Now we have that
$$2x^4 - 15x^3 + 16x^2 + 93x - 180 = (x - 3)^2(2x^2 - 3x - 20)$$

▶ Solve for the quadratic zeroes (use the quadratic formula or factor) to determine the remaining zeroes are $x = 4$ and $x = \dfrac{-5}{2}$.

I don't believe that anyone would argue this is a lot of work to do for just one problem! However, appreciate the power these two statements add to the tools we had for solving polynomial equations before technology became part of the process.

Let's do one more for the practice it gives.

EXAMPLE

▶ Solve the equation $6x^5 - 3x^4 - 4x^3 - 2x^2 - 2x + 5 = 0$.

▶ The factors of 6 are 1, 2, 3, and 6, while the factors of 5 are 1 and 5.

▶ Therefore, the set of possible rational roots is $\dfrac{p}{q} = \pm\left\{1, 5, \dfrac{1}{2}, \dfrac{5}{2}, \dfrac{1}{3}, \dfrac{5}{3}, \dfrac{1}{6}, \dfrac{5}{6}\right\}$.

▶ Descartes Rule of Signs tells us that there are 2 or 0 positive real zeroes and 3 or 1 negative real zeros (from $-6x^5 - 3x^4 + 4x^3 - 2x^2 + 2x + 5$).

▶ Try $x = 1$: $6 - 3 - 4 - 2 - 2 + 5 = 0$. Excellent, $x = 1$ is a zero. Reduce the polynomial.

$$
\begin{array}{r|rrrrrr}
1] & 6 & -3 & -4 & -2 & -2 & 5 \\
   &   & 6 & 3 & -1 & -3 & -5 \\
\hline
   & 6 & 3 & -1 & -3 & -5 & 0
\end{array}
$$

▶ We have
$6x^5 - 3x^4 - 4x^3 - 2x^2 - 2x + 5 = (x - 1)(6x^4 + 3x^3 - x^2 - 3x - 5)$

▶ Try $x = 1$ on the reduced polynomial: $6 + 3 - 1 - 3 - 5 = 0$ so $x = 1$ is again a double root.

$$
\begin{array}{r|rrrrr}
1] & 6 & 3 & -1 & -3 & -5 \\
   &   & 6 & 9 & 8 & 5 \\
\hline
   & 6 & 9 & 8 & 5 & 0
\end{array}
$$

▶ We have
$6x^5 - 3x^4 - 4x^3 - 2x^2 - 2x + 5 = (x - 1)^2(6x^3 + 9x^2 + 8x + 5)$.

▶ There is no need to see if $x = 1$ is a triple root because all the coefficients in the cubic polynomial are positive.

▶ Let's try $x = -1$: $-6 + 9 - 8 + 5 = 0$. We know $x = -1$ is a zero of the polynomial. Reduce the cubic.

$$
\begin{array}{r|rrrr}
-1 & 6 & 9 & 8 & 5 \\
 & & -6 & -3 & -5 \\
\hline
 & 6 & 3 & 5 & 0
\end{array}
$$

▶ Now we have
$$6x^5 - 3x^4 - 4x^3 - 2x^2 - 2x + 5 = (x-1)^2(x+1)(6x^2 + 3x + 5).$$

▶ Solve for the zeroes of the quadratic to find they are the complex numbers $\dfrac{-1}{4} \pm \dfrac{\sqrt{111}}{12}i$.

Maximum and Minimum Values of Polynomial Functions

We can now determine the end behavior of the polynomial, and we may be able to determine the zeroes of the polynomial with or without a calculator. The aspect of the graph's behavior that needs to be examined are the turning points. This topic is usually first taught in calculus. However, the problem in a strictly algebraic approach to calculus is that you need to be careful as to what functions you choose so that an algebraic solution can be found.

Using graphing technology to graph $f(x) = 2x^4 - 15x^3 + 16x^2 + 93x - 180$, we see the graph when we limit the domain of the window from -5 to 5 and

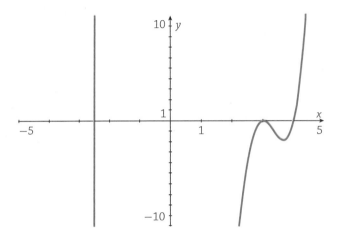

the range of the window from –10 to 10. When we expand the range of the window to the interval –300 to 300, we get a different look.

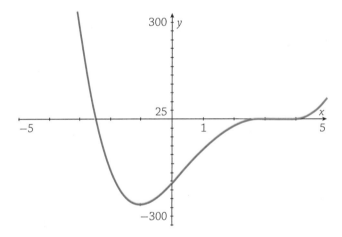

The graph has a lower turning point (which we now call a **local minimum**—it is the smallest value relative to all the other points near it but not necessarily the smallest value of the function). We can use the

minimum command in the graphing technology to see the minimum value is –240 and occurs when x is approximately –1.05. (The **maximum command** is used to determine the local maximum value.)

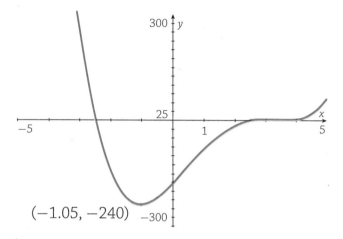

Looking back at the first view of the graph, we see that 0 is a local maximum and –1.83 is an approximation for the local minimum.

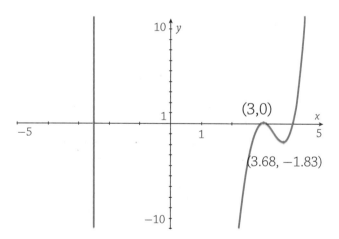

Together, the local minimum and local maximum values are called the **local extrema** of the function.

▶ Determine the values of the local extrema for the function $f(x) = x^4 - 5x^2 + 3$.

▶ Use the graphing technology to determine the graph has a local maximum of 3 at $x = 0$ and local minimum of −3.25 at $x = \pm 1.58$.

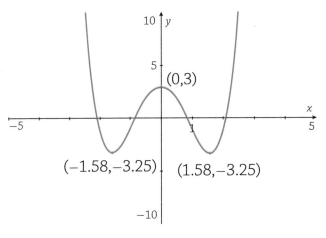

Here is an example of an application for this concept.

▶ The number of miles a delivery truck is from the distribution center on a particular day is given by the function
$$f(x) = \frac{-1}{100}x^5 + \frac{11}{100}x^4 - \frac{9}{25}x^3 + \frac{18}{25}x^2 - \frac{32}{25}x + \frac{53}{25}, \text{ where } x$$
represents the number of hours passed after 9 a.m.

▶ Determine when the truck is closest and farthest from the distribution center between 9 a.m. and 3 p.m.

▶ The interval from 9 a.m. to 3 p.m. corresponds to the interval $x = 0$ to $x = 6$. Sketch the graph on that interval.

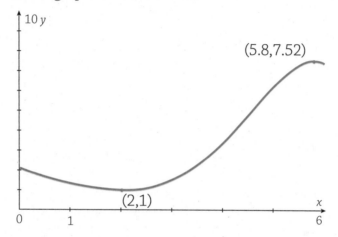

▶ The truck is 1 mile from the distribution center at 11 a.m. and 7.52 miles from the distribution center at 2:48 p.m.

EXERCISES

EXERCISE 3-1

Answer each of the following polynomial-related questions.

1. Factor $128x^7 - 1$.

2. Determine the remainder when $2x^5 - 3x^4 + 7x^2 + 5x - 11$ is divided by $x - 1$.

3. Is $x - 1$ a factor of $2x^5 - 3x^4 + 7x^2 + 5x - 11$? Explain.

4. What is the quotient when $2x^5 - 3x^4 + 7x^2 + 5x - 11$ is divided by $x - 1$?

5. Use synthetic division to divide $6x^4 + x^3 - 39x^2 + 6x + 40$ by $3x - 4$.

6. Use synthetic division to divide $2x^3 + 3x^2 - 9x - 10$ by $2x + 5$.

7. Completely factor $6x^4 + x^3 - 39x^2 + 6x + 40$.

8. Determine the end behavior of the function $f(x) = 8x^5 - 12x^3 + 4x^2 + 9$.

9. Determine the end behavior of the function $g(x) = -12x^6 + 41x^4 + 10x^2 + 7x$.

10. Determine the end behavior of the function $p(x) = -8x^9 + 102x^6 - 40x^3 + 9$.

EXERCISE 3-2

Factor each of the polynomials in questions 1 to 5.

1. $4x^4 - 16x^3 - 147x^2 + 245x + 1372$

2. $x^5 - 7x^4 - 2x^3 + 100x^2 - 232x + 160$

3. $6x^4 + 17x^3 + 25x^2 + 34x - 40$

4. $x^6 - 4x^4 - 16x^2 + 64$

5. $8x^5 + 46x^4 + 89x^3 + 52x^2 - 63x - 36$

Solve each of the equations in questions 6 to 10.

6. $4x^4 - 16x^3 - 147x^2 + 245x + 1372 = 0$

7. $x^5 - 7x^4 - 2x^3 + 100x^2 - 232x + 160 = 0$

8. $6x^4 + 17x^3 + 25x^2 + 34x - 40 = 0$

9. $x^6 - 4x^4 - 16x^2 + 64 = 0$

10. $8x^5 + 46x^4 + 89x^3 + 52x^2 - 63x - 36 = 0$

Solve each of the inequalities in questions 11 to 15, but don't forget question 16!

11. $4x^4 - 16x^3 - 147x^2 + 245x + 1372 < 0$

12. $x^5 - 7x^4 - 2x^3 + 100x^2 - 232x + 160 \geq 0$

13. $6x^4 + 17x^3 + 25x^2 + 34x - 40 \leq 0$

14. $x^6 - 4x^4 - 16x^2 + 64 \geq 0$

15. $8x^5 + 46x^4 + 89x^3 + 52x^2 - 63x - 36 > 0$

16. Use the Rational Root theorem to determine the set of all possible rational roots for the polynomial $30x^3 + 83x^2 + 41x - 24$.

Determine the values of the local extrema for each of the functions in questions 17 to 20.

17. $f(x) = 4x^5 - 12x^3 + x^2 - 2$

18. $p(x) = 5x^4 - 8x^3 - x + 2$

19. $r(x) = -2x^3 + 6x^2 + 4$

20. $q(x) = -x^4 + 12x^2 + 3$

Flashcard App

4 Rational Functions

MUST KNOW

 Discontinuities of rational functions are limited to vertical asymptotes and point discontinuities.

 Reasonably good sketches of rational functions can be made based on the knowledge of intercepts, discontinuities, and end behavior.

olynomial functions have the important characteristic that they are continuous. Continuity is a topic that you will study in depth in your first semester of Calculus. For now, accept this as a working definition of a continuous function—a function that can be drawn without ever having to lift your pencil off the page. Some rational functions are continuous. However, most are not. In this chapter we will look at both types of rational functions.

Intercepts for the Graphs of Rational Functions

How do you determine the value of the y-intercept for the graph of a rational function? As always, you set x equal to zero and evaluate the function. If $x = 0$ is not in the domain of the function, the graph will not have a y-intercept. You find the values of the x-intercepts as you always have—set the function equal to zero and solve. If the equation has no solutions, then the graph never crosses the x-axis.

EXAMPLE

▸ Determine the values of the x- and y-intercepts for the function $f(x) = \dfrac{x^2 - 4x - 5}{x^2 - 4}$.

▸ The y-intercept is found at $f(0)$ and equals $\dfrac{5}{4}$. The x-intercepts are found by solving $f(x) = 0$, and the solution to this is $x = 5, -1$.

The solution to that problem seems to be very easy. Let's look at another seemingly easy problem that really isn't.

▶ Determine the values of the x- and y-intercepts for the function
$f(x) = \dfrac{x^2 - 4x - 5}{x^2 - 1}$.

▶ The y-intercept is found at $f(0)$ and equals 5. The x-intercepts are found by solving $f(x) = 0$, and the solution to this is $x = 5, -1$. This seems to be fairly straightforward except for the fact that -1 is not in the domain of the problem.

▶ The only x-intercept for this function is at $x = 5$.

This second example gives us a chance to remember an important rule: Identifying the domain of the function is the first step when dealing with rational (and irrational) functions.

Discontinuities of Rational Functions

A discontinuity is a break in the graph of a function. There are three types of discontinuities: jump, point, and infinite. You very likely saw a jump discontinuity when you studied piecewise functions. For example, the

function $f(x) = \begin{cases} x^2 & x < 0 \\ x + 1 & x \geq 0 \end{cases}$ has a jump discontinuity at $x = 0$.

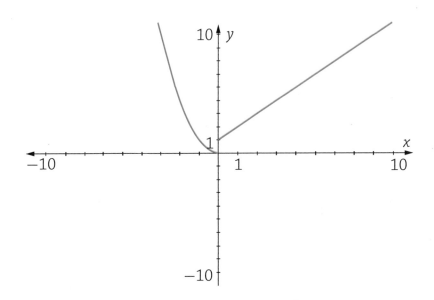

There will always be a jump from the point (0, 0) from the left branch of this graph to the point (0, 1) on the right branch.

The function $f(x) = \dfrac{x^2 - 4x - 5}{x^2 - 1}$ has a point discontinuity. Point discontinuities occur when there is a common factor in the numerator and denominator of the rule defining the function. In this case,

$f(x) = \dfrac{(x+1)(x-5)}{(x+1)(x-1)} = \dfrac{x-5}{x-1}$. The point discontinuity will occur at the point (−1, 3). Point discontinuities are difficult to see when you graph the function on your device. You will know the discontinuity is there when you look at a table of values or ask the device to evaluate $f(-1)$.

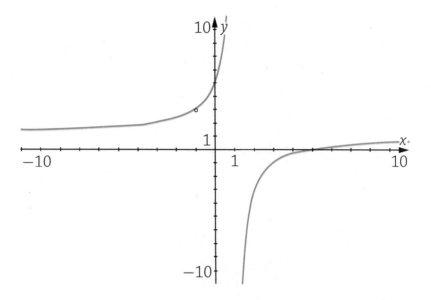

Infinite discontinuities take on one of three forms: one branch goes to negative infinity while the other goes to positive infinity, both branches go to negative infinity, or both branches go to positive infinity.

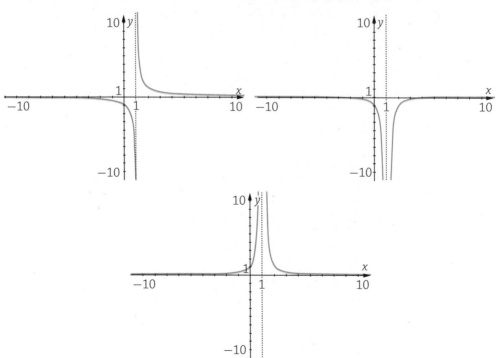

Being able to identify the type of discontinuity a rational has, if it has one, is a useful skill in the field of analysis (calculus). This characteristic, with the other topics in this chapter, helps form a checklist to assist in sketching rational functions. Before you get to "my graphing calculator can do this," keep in mind it is not unusual to be given the behaviors of the function without being given the rule for the function. Consequently, this material might be more pertinent than you originally thought.

What, then, should we be looking for? It is easy to determine the points of discontinuity with rational functions because these are the values that cause the denominator to be zero. If, for example, $x = 3$ causes the denominator to equal zero, check to see if $x - 3$ is a factor of the numerator. If it is, you have a potential point discontinuity. (Did you notice I used the word "potential"? More on that soon.) If there is no common factor, then there is an infinite discontinuity. But which type is it? Signs analysis will help with that. Check the sign of the rational expression to the left and right of the point of discontinuity.

EXAMPLE

▶ What type of discontinuity does the function $f(x) = \dfrac{x^2 + 5x + 6}{x + 3}$ exhibit?

▶ The numerator factors to $(x + 3)(x + 2)$, so the expression reduces to $x + 2$; this function therefore has a point discontinuity. Substituting -3 for x shows that the discontinuity will occur at the point $(-3, -1)$.

Let's look at the "potential" point of discontinuity now.

EXAMPLE

▶ What type of discontinuity does the function $g(x) = \dfrac{x^2 - 9}{(x - 3)^2}$ exhibit?

▶ Clearly $x = 3$ is a value for which there will be a discontinuity, and $x - 3$ is a factor of both the numerator and denominator. However, when the expression is simplified, there is still a factor of $x - 3$ in the denominator. So, unlike the last example, we cannot simply substitute $x = -3$ into the reduced expression to determine the point discontinuity.

▶ Using the reduced expression, $g(x) = \dfrac{x+3}{x-3}$, we see that for values smaller than 3, the numerator is positive while the denominator is negative, so the graph goes to negative infinity as x approaches 3 from the left. For values of x just larger than 3, we see that both the numerator and denominator are positive.

> **BTW**
>
> *Analyzing functions from the left- and right-hand sides of a given value happens fairly often. The phrase "the graph goes to negative infinity as x approaches 3 from the left" has the notation $x \to 3^- \Rightarrow y \to -\infty$, while the phrase "the graph goes to positive infinity as x approaches 3 from the right" has the notation $x \to 3^+ \Rightarrow y \to \infty$. The negative and positive "exponents" merely indicate the direction from which the number is being approached.*

▶ We conclude that the graph goes to positive infinity as x approaches 3 from the right.

The following example also looks at infinite discontinuities.

▶ What type of discontinuity does the function $r(t) = \dfrac{t-9}{(t-3)^2}$ exhibit?

(I chose t as the variable because every once in a while I like to mix it up.)

▶ We see that, once again, there is a discontinuity at 3. Analysis shows $t \to 3^- \Rightarrow y \to -\infty$ and $t \to 3^+ \Rightarrow y \to -\infty$. Consequently, we know that both branches of this graph go to negative infinity as t approaches 3, and we write $t \to 3 \Rightarrow y \to -\infty$. (The absence of an exponent indicates that the graph exhibits the same behavior on either side of the point in question.)

The next problem is a bit more complicated due to the number of factors involved. We'll use signs analysis to help us examine the discontinuities.

▶ What type of discontinuity does the function $f(x) = \dfrac{x^2 - 3x - 18}{x^2 + x - 30} =$

$\dfrac{(x-6)(x+3)}{(x-5)(x+6)}$ exhibit?

▶ When we set up the number line to do the signs analysis, we'll include both the zeroes of the function and the points of discontinuity. We do this so that we can account for each sign change. As we did in the last chapter, we'll include an "$= 0$" for the zeroes of the function. We'll use a vertical dotted line (an asymptote) for each of the points of discontinuity.

$$\dfrac{(-)(-)}{(-)(-)} \to + \qquad \dfrac{(-)(-)}{(-)(+)} \to - \; = 0 \quad \dfrac{(-)(+)}{(-)(+)} \to + \qquad \dfrac{(-)(+)}{(+)(+)} \to - \; = 0 \quad \dfrac{(+)(+)}{(+)(+)} \to +$$

$x = -7$	$x = -4$	$x = 0$	$x = 5.5$	$z = 10$
-6	-3	5	6	

▶ Since all we are concerned about in this problem is the nature of the discontinuities, we'll describe the behavior of the function near $x = -6$ and $x = 5$.

▶ $x \to -6^- \Rightarrow y \to \infty$ and $x \to -6^+ \Rightarrow y \to -\infty$;
$x \to 5^- \Rightarrow y \to -\infty$ and $x \to 5^+ \Rightarrow y \to \infty$

End Behavior for Rational Functions

As we did with polynomials, we need to look at the end behavior of rational functions. This turns out to be fairly simple. Let $f(x) = \dfrac{g(x)}{k(x)}$ where both

$g(x)$ and $k(x)$ are polynomial functions. One of three conditions must be true:

- The degrees of $g(x)$ and $k(x)$ are equal.

- The degree of $g(x)$ is greater than the degree of $k(x)$.

- The degree of $g(x)$ is less than the degree of $k(x)$.

If the degrees of $g(x)$ and $k(x)$ are equal, then as $x \to \infty$, $f(x)$ will approach the ratio of the coefficients of the terms of highest degree in both the numerator and denominator.

EXAMPLE

▶ Describe the end behavior of the function $f(x) = \dfrac{4x^3 - 8x^2 + 7x + 2}{8x^3 + 3x - 6}$.

▶ To understand why the solution is the ratio of the coefficients of highest degree, divide the numerator and denominator by x^3,

$f(x) = \dfrac{4 - \dfrac{8}{x} + \dfrac{7}{x^2} + \dfrac{2}{x^3}}{8 + \dfrac{3}{x^2} - \dfrac{6}{x^3}}$. As $x \to \infty$ (or as $x \to -\infty$), all the terms

except the 4 and 8 will go to zero. (This makes sense—a small number divided by an infinitely large number will be negligibly small.)

▶ Therefore, as $x \to \infty$, $f(x) \to \dfrac{1}{2}$. The line $y = \dfrac{1}{2}$ is called a **horizontal asymptote**.

If the degree of the numerator is greater than the degree of the denominator, the end behavior of the graph will be to go to one of the infinities, depending on the lead coefficient of the terms of highest degree.

▶ Describe the end behavior of the function $f(x) = \dfrac{4x^3 - 8x^2 + 7x + 2}{8x^2 + 3x - 6}$.

▶ Divide both numerator and denominator by x^3, the term of highest

degree, $f(x) = \dfrac{4 - \dfrac{8}{x} + \dfrac{7}{x^2} + \dfrac{2}{x^3}}{\dfrac{8}{x} + \dfrac{3}{x^2} - \dfrac{6}{x^3}}$. As $x \to \infty$, the denominator gets

infinitely small, and 4 divided by an infinitely small number is an infinitely large number. The sign for the denominator will be positive because 8 divided by a positive number is still positive.

▶ Therefore, as $x \to \infty$, $f(x) \to \infty$. On the other hand, as $x \to -\infty$, $f(x) \to -\infty$.

Lastly, if the degree of the numerator is less than the degree of the denominator, the end behavior of the graph will be to go to zero.

▶ Describe the end behavior of the function $f(x) = \dfrac{4x^3 - 8x^2 + 7x + 2}{8x^4 + 3x - 6}$.

▶ Divide both numerator and denominator by x^4, the term of highest

degree, $f(x) = \dfrac{\dfrac{4}{x} - \dfrac{8}{x^2} + \dfrac{7}{x^3} + \dfrac{2}{x^4}}{8 + \dfrac{3}{x^3} - \dfrac{6}{x^4}}$.

▶ As $x \to \pm\infty$, the entire numerator goes to zero, and 0 divided by 8 is zero.

Graphing Rational Functions

Let's put all these pieces together—and a new piece that we'll add in this section—to get a feel for what the graph of some rational functions will look like.

EXAMPLE

▶ Analyze and sketch a graph of $f(x) = \dfrac{2x - 4}{x^2 - 16}$.

▶ The y-intercept is at $\left(0, \dfrac{1}{4}\right)$, while the x-intercept is at $(2, 0)$. There are discontinuities at $x = \pm 4$. A signs analysis shows the behavior around the discontinuities.

$$\frac{(-)}{(-)(-)} \to - \qquad \frac{(-)}{(+)(-)} \to + \qquad \frac{(+)}{(+)(-)} \to - \qquad \frac{(+)}{(+)(+)} \to +$$
$$= 0$$

| $x = -5$ | -4 | $x = -2$ | 2 | $z = 3$ | 4 | $z = 5$ |

▶ The degree of the numerator is less than the degree of the denominator, as $x \to \pm\infty$, $f(x) \to 0$.

▶ We know the graph begins below the x-axis, and as it moves toward the vertical asymptote at $x = -4$, the values grow infinitely large, but negative. To the right of the vertical asymptote at $x = -4$, the functional values are positive but dropping. The graph crosses the y-axis at one-fourth and then the x-axis at two and continues to drop as the graph approaches the vertical asymptote at $x = 4$. The values of the function are positive to the right of the asymptote at $x = 4$ but decrease as the values of x increase. Finally, as the values of x get infinitely large, the values of y approach zero.

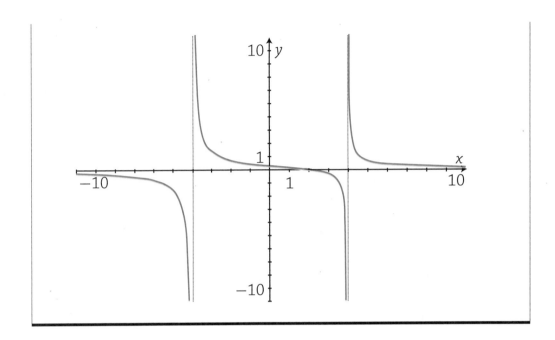

That wasn't too bad. Let's look at the graph of a function that has a slight change from the last graph.

EXAMPLE

▶ Analyze and sketch the graph of the function $f(x) = \dfrac{2x^2 - 8}{x^2 - 1} =$

$\dfrac{2(x+2)(x-2)}{(x+1)(x-1)}$.

▶ The intercepts are at $(0, 8)$, $(-2, 0)$, and $(2, 0)$. The degree of the numerator and denominator are equal, so the graph will have a horizontal asymptote at $y = 2$. A signs analysis shows

$$\frac{(-)(-)}{(-)(-)} \to + \quad = 0 \qquad \frac{(+)(-)}{(-)(-)} \to - \qquad \frac{(+)(-)}{(+)(-)} \to + \qquad \frac{(+)(-)}{(+)(+)} \to - \quad = 0 \qquad \frac{(+)(+)}{(+)(+)} \to +$$

| $x = -3$ | -2 | $x = -\dfrac{3}{2}$ | -1 | $x = 0$ | 1 | $x = \dfrac{3}{2}$ | 2 | $x = 3$ |

▶ The graph follows the horizontal asymptote $y = 2$ from below the asymptote, decreasing all the while, crosses the x axis at $x = -2$, and then follows the vertical asymptote at $x = -1$. The graph has positive values between the vertical asymptotes $x = -1$ and $x = 1$ and has a y-intercept at $y = 8$. The graph rises from the right of the vertical asymptote $x = 1$, crosses the x-axis at $x = 2$, and then follows the horizontal asymptote $y = 2$ as x goes to infinity.

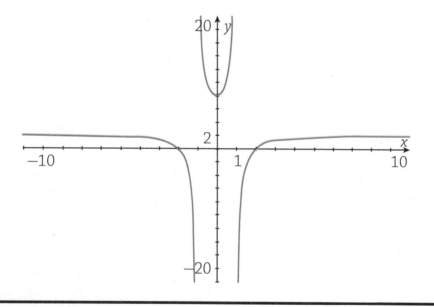

Let's take a look at another example.

EXAMPLE

▶ Analyze and sketch a graph of the function $f(x) = \dfrac{2x^2 + 9x - 5}{x - 2}$.

▶ A quadratic polynomial divided by a linear expression will yield a linear expression with or without a remainder. Use synthetic division to examine this new function.

$$\begin{array}{r|rrr} 2 & 2 & 9 & -5 \\ & & 4 & 26 \\ \hline & 2 & 13 & 21 \end{array}$$

▶ We know then that $f(x) = \dfrac{2x^2 + 9x - 5}{x - 2} = 2x + 13 + \dfrac{21}{x - 2}$.

▶ There will be a vertical asymptote at $x = 2$, the y-intercept is $\dfrac{5}{2}$, and the x-intercepts are at -5 and $\dfrac{1}{2}$. As for end behavior, as $x \to \pm\infty$, $\dfrac{21}{x - 2} \to 0$, so the graph will follow the line $y = 2x + 13$. (This line is called an oblique asymptote.) The sketch of the graph is as shown.

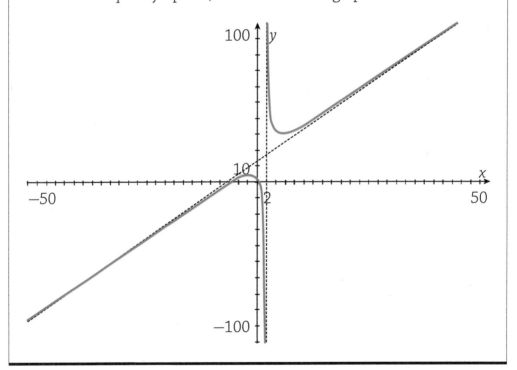

Maxima and Minima of Rational Functions

As we saw in the last section, the graph of the function $f(x) = \dfrac{2x^2 - 9x - 5}{x - 2}$ has both a local maxima and minima. They occur at the points with the approximate coordinates $(-1.24, 4.04)$ and $(5.24, 30)$. (It turns out that the exact values of x for these points are $\dfrac{4 \pm \sqrt{42}}{2}$. We need some knowledge of calculus to determine these values.)

In Chapter 2, we solved the problem involving Brendon and Colin's plans to build a garden. This next problem is similar in application but different in the mathematics.

EXAMPLE

▶ Brendon and Colin plan to build a rectangular garden with an area of 1200 square feet. One edge of the garden is along the creek bed and requires a stronger fence than do the other three sides. When Brendon and Colin go to the store to buy fencing, they find that the fencing along the creek bed costs $4 per foot, while the fencing for the remaining sides costs $3 per foot. What are the dimensions of the garden they can build for a minimum cost?

▶ A diagram of the problem shows the creek and the remaining sides.

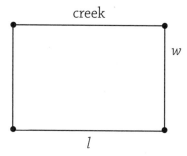

creek

w

l

▶ We know the area is given by the equation $lw = 1200$, so $l = \dfrac{1200}{w}$,
and the cost of the fencing is $C = 4l + 3(l + 2w) = 7l + 2w$. Rewriting
the cost equation in terms of w, $C = 7\left(\dfrac{1200}{w}\right) + 2w = \dfrac{8400}{w} + 2w$.

▶ An analysis of the sketch of the graph of the cost function (using
$w > 0$) shows that when the width of the garden is 64.8 feet long, the
cost of the garden will be approximately $259 (actually, $259.23—
try it on your graphing device). The dimensions of the garden with
minimum cost will be 64.8 feet by 18.52 feet.

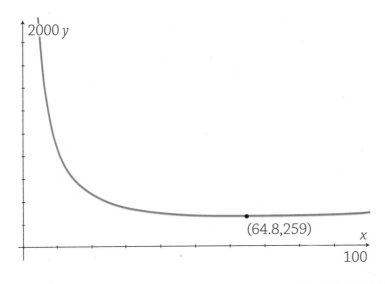

(64.8,259)

This next problem is also geometric in nature but requires a rational
function in its solution.

EXAMPLE

▶ The Sales Soup Company packages its soups in cylindrical cans with
a volume of 16 ounces (473.18 cubic centimeters). Determine the
dimensions of the cylindrical cans that have a minimum surface area.

▶ The volume of a cylinder is given by the formula $V = \pi r^2 h$, where r is the radius of the base and h is the height of the cylinder. In this example, $\pi r^2 h = 473.18$. The surface area of the cylinder is given by the formula $A = 2\pi r^2 + 2\pi r h$. Solving for h in the volume formula gives $h = \dfrac{473.18}{\pi r^2}$.

▶ Writing the area formula in terms of the radius of the base yields $A = 2\pi r^2 + 2\pi r\left(\dfrac{473.18}{\pi r^2}\right) = 2\pi r^2 + \dfrac{946.36}{r}$. Analysis of the graph of this function shows that the minimum surface area is achieved when the radius is 4.22 cm. Consequently, the height of the cylinder will be 8.45 cm.

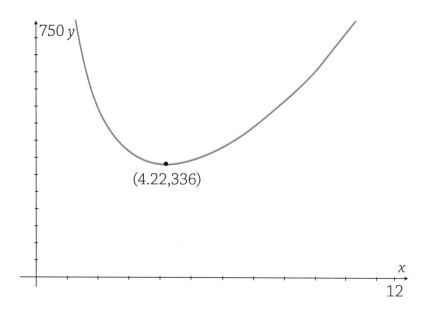
(4.22,336)

EXERCISES

EXERCISE 4-1

Determine the coordinates of the intercepts in questions 1 to 5.

1. $b(z) = \dfrac{z^2 - 8z - 9}{z^2 - 9}$

2. $c(v) = \dfrac{v + 5}{v^2 + 1}$

3. $d(p) = \dfrac{4p^2 - 4p + 1}{4p^2 - 1}$

4. $e(t) = \dfrac{t^3 - 4t}{t^2 - 25}$

5. $w(a) = \dfrac{6a - 12}{a^2 - 4}$

Determine the nature of any discontinuities for the functions in questions 6 to 10.

6. $b(z) = \dfrac{z^2 - 8z - 9}{z^2 - 9}$

7. $c(v) = \dfrac{v + 5}{v^2 + 1}$

8. $d(p) = \dfrac{4p^2 - 4p + 1}{4p^2 - 1}$

9. $e(t) = \dfrac{t^3 - 4t}{t^2 - 25}$

10. $w(a) = \dfrac{6a - 12}{a^2 - 4}$

EXERCISE 4-2

Describe the end behavior for each function defined in questions 1 to 5.

1. $f(x) = \dfrac{x^2 - 25}{x^2 - 4}$

2. $g(x) = \dfrac{2x^2 - x - 6}{x^2 + 3x - 4}$

3. $h(x) = \dfrac{2x^2 - x - 15}{3x^2 - 2x - 8}$

4. $k(x) = \dfrac{x^4 + 2x^3 - 9x - 2}{x^5 + x^4 + 10}$

5. $m(x) = \dfrac{x^3 - 4x - 6}{x^2 + 5x + 12}$

Analyze and graph the functions defined in questions 6 to 10.

6. $f(x) = \dfrac{x^2 - 25}{x^2 - 4}$

7. $g(x) = \dfrac{2x^2 - x - 6}{x^2 + 3x - 4}$

8. $h(x) = \dfrac{2x^2 - x - 15}{3x^2 - 2x - 8}$

9. $p(x) = \dfrac{x^2 - 3x + 4}{x - 2}$

10. $r(x) = \dfrac{(x - 1)^2(x^2 + 16)}{x^4 - 16}$

Determine the values of the local extrema in questions 11 to 13.

11. $f(x) = \dfrac{5x}{x^2 + 4}$

12. $p(x) = \dfrac{x^2 - 3x + 4}{x - 2}$

13. $r(x) = \dfrac{(x - 1)^2(x^2 + 16)}{x^4 - 16}$

Solve the problems for the extreme solutions in questions 14 and 15.

14. The Sales Soup Company packages its soups in cylindrical cans with a volume of 16 ounces (473.18 cubic centimeters). If the material forming the top of the cylinder needs to be from a material that costs $0.02 per cm while the rest of the cylinder can be made from a material that costs $0.015 per cm, determine the dimensions of the cylinders that can be made for a minimum cost.

15. Carson and Cameron plan to build a rectangular garden with an area of 1800 square feet. The edge of the garden is along the creek bed and requires a stronger fence than do the other three sides. When Carson and Cameron go to the store to buy fencing, they find that the fencing along the creek bed costs $6 per foot, while the fencing for the remaining sides costs $4 per foot. What are the dimensions of the garden they can build for a minimum cost?

Flashcard App

Conic Sections

MUST KNOW

 Conic sections can be described by a single number, the eccentricity.

 Conic sections are used in the design of television reception, stereo speakers, headlights, and brass instruments.

I am about to do something I swore that I would never, ever do: write a sentence beginning with "When I was younger." Considering that I am in my mid-60s and I don't feel like 30 years ago was all that long ago, I have a newfound understanding of what my elders used to say.

When I was younger, there were many different authors of calculus books. However, the names of the books were all the same—Calculus with Analytic Geometry. The bulk of the geometry discussed was the conic sections. During my career as a high school mathematics teacher, the topic of the conic sections moved from Precalculus to Algebra II to the back of the closet. The mathematical reforms during the last 25 to 30 years have emphasized numerical analysis and technology, so many of the older topics have been put aside. Given all that, we will look at the conic sections from both a geometrical and numerical approach.

We do not know much more about the conic sections than did the ancient Greeks. That is not to say that the conic sections are not worth knowing, but rather the eight-volume series written by Apollonius was that complete. He began with two cones that shared a common vertex and whose bases were parallel to each other.

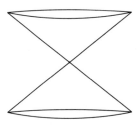

He then had a plane cut through this figure and analyzed the results. The three basic cross sections (the intersection of the plane with the double cone) are a single point, a single line, and intersecting lines.

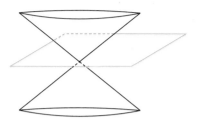 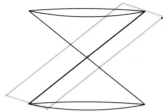 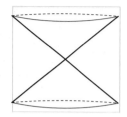

The plane is parallel to the bases and passes through the common vertex.

The plane is parallel to one edge of the cone and contains the edge. (That is, the plane is tangent to the cone.)

The plane is perpendicular to the two bases of the cone and contains the common vertex of the cones.

The other intersections he considered are displayed in each of the following sections.

Analyze and Graph Circles

At this point in your study of mathematics, there isn't much more that can be said about the circle than you already know. The geometric definition that Apollonius provided is that the circle is the cross section of the two cones when the plane is parallel to the base but does not contain the common vertex of the two cones.

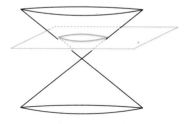

The analytic definition for the circle is the set of points, or **locus** of points, in a plane that are equidistant from a fixed point. The fixed point, as you know, is the center of the circle and the distance from the center to any point on the circle is called the length of the radius.

> ▶ Determine the equation for the locus of points 8 units from the point (−3, 5).
>
> ▶ Using (x, y) to represent an arbitrary point on the locus, we use the distance formula to write $\sqrt{(x - (-3))^2 + (y - 5)^2} = 8$.
>
> ▶ Manipulating this equation, we get $(x + 3)^2 + (y - 5)^2 = 64$.

Analyze and Graph Parabolas

When the plane is parallel to a side of the cone and passes through the cone, the resulting figure is a **parabola**.

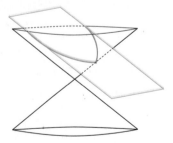

The analytic definition of the parabola is that it is the locus of points equidistant from a fixed point, the **focus**, and a fixed line, the **directrix**.

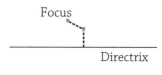

Focus

Directrix

Focus

Directrix

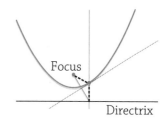

Focus

Directrix

One point equidistant from the focus and directrix.

The perpendicular bisector of the segment joining the focus to a point on the directrix.

The locus of all the points equidistant from the focus and directrix.

For the sake of making the expression easier to manipulate, the coordinates of the focus are $(0, p)$ and the equation of the directrix is $y = -p$. Again, the coordinates of an arbitrary point, P, on the locus is (x,y). The distance from the focus to P is $\sqrt{x^2 + (y - p)^2}$, while the distance from P to the directrix is $y + p$. Setting the expressions equal to each other and manipulating them, we get $\sqrt{x^2 + (y - p)^2} = y + p$, which becomes $x^2 + (y - p)^2 = (y + p)^2$. After we expand the square binomials, gather terms, and simplify, this equation becomes $y = \dfrac{1}{4p}x^2$. You know this parabola has its vertex at the origin. In fact, the point midway between the focus and the directrix is called the **vertex** of the parabola. Also, the line perpendicular to the directrix and passing through the focus is the **axis of symmetry**.

BTW

If the coordinates of the focus are (p,0) and the equation of the directrix is x = -p, the equation of the parabola formed is $x = \dfrac{1}{4p}y^2$, the parabola that opens horizontally.

▶ The basic parabola that you have manipulated since Algebra I is $y = x^2$. What are the coordinates of the focus and the equation of the directrix?

▶ In this case, the coefficient 1 must be equal to the coefficient $\dfrac{1}{4p}$.

Consequently, $p = \dfrac{1}{4}$.

▶ The coordinates of the focus are $\left(0, \dfrac{1}{4}\right)$ and the equation for the directrix is $y = \dfrac{-1}{4}$.

We know from our work in Algebra II that the equation of the parabola with vertex (h, k) is $y - k = a(x - h)^2$. So now we know that the leading coefficient under the former way we wrote the equation of the parabola is related to the length of the segment from the vertex to the focus (or the vertex to the directrix).

Let's take a look at a few problems that follow from this locus definition for the parabola.

▶ Determine the equation of the parabola whose focus has coordinates $(5, 8)$ and whose directrix has equation $y = 12$.

▶ The distance from the focus to the directrix is $2p$. In this case, $2p = 4$ so $p = 2$. However, the directrix is above the focus, so we assign a direction to this value and $p = -2$. The coordinates of the vertex for this parabola are $(5, 10)$.

▶ Therefore, the equation of the parabola is $y - 10 = \dfrac{-1}{8}(x - 5)^2$.

This next example shows an interesting characteristic of the parabola.

EXAMPLE

▶ Determine the length of the segment that connects two points on the parabola $y - 10 = \dfrac{-1}{8}(x - 5)^2$, is parallel to the directrix, and passes through the focus.

▶ The segment in question must be part of the horizontal line $y = 8$. Substituting this value into the equation,

$$8 - 10 = \frac{-1}{8}(x - 5)^2 \Rightarrow -2 = \frac{-1}{8}(x - 5)^2 \Rightarrow 16 = (x - 5)^2 \Rightarrow x = 1, 9$$

▶ The distance between these two points is 8, the value of $|4p|$. Is that a coincidence? No, it's not!

▶ The chord with endpoints on the parabola and that passes through the focus is called the **focal chord** (though it used to be known as the latus rectum). The length of the focal chord is $|4p|$.

Let's do one more example of the parabola before we talk about the reflective properties of this locus.

EXAMPLE

▶ Determine the coordinate of the vertex, the coordinates of the focus, the equation of the directrix, and the length of the focal chord of the parabola $x + 2 = \dfrac{-1}{20}(y - 6)^2$.

▶ We can read the coordinates of the vertex, $(-2, 6)$, and the length of the focal chord, 20, directly from the equation. We can also determine the value of p is -5, making the focus to the left of the vertex and the directrix to the right.

▶ The coordinates of the focus are (−7, 6), and the equation of the directrix is $x = 3$.

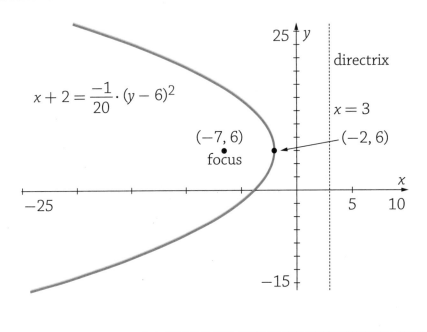

The parabola has a very useful characteristic that any ray that intersects the interior of the parabola will reflect to pass through the focus. In addition, any ray that passes through the focus will reflect in such a way that it is perpendicular to the directrix. This is based on the physical property that *the angle of incidence is equal to the angle of reflection.*

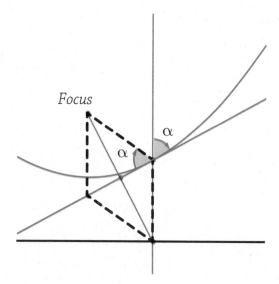

People who have satellite TV in their homes use the parabolic dish to receive their signal. The actual receiver is the focus of the dish. You might also see the parabolic dishes on the sidelines of outdoor sporting events as the network attempts to hear what is being said on the field.

Headlights and spotlights also use this reflective property. The bulb is at the focus and is shown into the housing to reflect off a parabolic mirror, thus concentrating the light rays so that they do not dissipate much beyond the source.

Analyzing and Graphing Ellipses

When a plane cuts a cone at an angle that is not parallel to the edge of the cone but not so steep as to intersect the second nap of the cone, the resulting cross section is an **ellipse**.

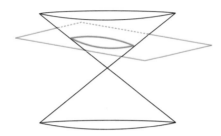

The analytic definition for an ellipse is that it is the locus of points from two fixed points so that the sum of the distances is a constant. The traditional derivation of the ellipse based on this statement might make you wonder how the constants were selected. I have no good answer for you but caution you to pay attention!

EXAMPLE

▶ Identify the fixed points (**foci**) with the coordinates $(-c, 0)$ and $(c, 0)$ (thus making the origin the midpoint of the segment joining the two foci). In addition, identify the fixed sum of the distances by $2a$.

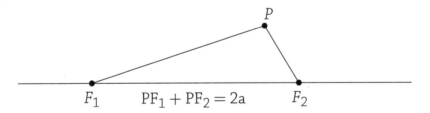

▶ This statement generates the equation $\sqrt{(x + c)^2 + y^2} + \sqrt{(x - c)^2 + y^2} = 2a$. Bear with me here as the algebra gets a bit ugly. In order to simplify this equation, we'll need to isolate one of the radical statements: $\sqrt{(x + c)^2 + y^2} = 2a - \sqrt{(x - c)^2 + y^2}$.

▸ Square both sides of the equation: $(x+c)^2 + y^2 =$

$$\left(2a - \sqrt{(x-c)^2 + y^2}\right)^2 = 4a^2 - 4a\sqrt{(x-c)^2 + y^2} + (x-c)^2 + y^2$$

▸ Subtract y^2 from both sides of the equation and expand the square trinomials.

$$x^2 + 2cx + c^2 = 4a^2 - 4a\sqrt{(x-c)^2 + y^2} + x^2 - 2cx + c^2$$

▸ Bring all the terms outside the radical to the left side of the equation.

$$4cx - 4a^2 = -4a\sqrt{(x-c)^2 + y^2} \Rightarrow a^2 - cx = a\sqrt{(x-c)^2 + y^2}$$

▸ Square both sides of the equation: $(a^2 - cx)^2 = a^2((x-c)^2 + y^2)$

▸ Expand: $a^4 - 2a^2cx + c^2x^2 = a^2x^2 - 2a^2cx + a^2c^2 + a^2y^2$

▸ Combine like terms: $a^4 - a^2c^2 + c^2x^2 - a^2x^2 = a^2y^2 \Rightarrow a^2(a^2 - c^2) + (c^2 - a^2)x^2 = a^2y^2$

▸ Move the terms in x and y to the same side of the equation:

$a^2(a^2 - c^2) = (a^2 - c^2)x^2 + a^2y^2$

▸ Here comes the head scratcher. Let $b^2 = a^2 - c^2$: $a^2b^2 = b^2x^2 + a^2y^2$.

▸ Divide both sides of the equation by the constant a^2b^2: $\dfrac{x^2}{a^2} + \dfrac{y^2}{b^2} = 1$.

▸ We now have the equation of the ellipse with the center at the origin with the foci on the x-axis is $\dfrac{x^2}{a^2} + \dfrac{y^2}{b^2} = 1$.

▶ Let's go back to that head scratcher. Place point P on the perpendicular bisector of the segment joining the foci.

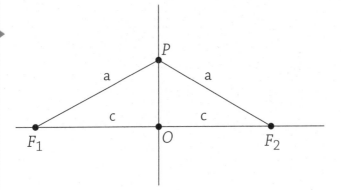

▶ We know from the Pythagorean theorem that $OP^2 + OF_2^2 = PF_2^2$ or that $OP^2 + c^2 = a^2$. It was just natural for the segment OP to be designated with the value b.

IRL If you ask anyone who has gotten past seventh grade to tell you what the Pythagorean theorem says, almost 100 percent will respond $a^2 + b^2 = c^2$, rather than the sum of the squares of the lengths of the two shorter sides of the right triangle is equal to the square of the length of the hypotenuse!

With $b^2 = a^2 - c^2$, we also get $c^2 = a^2 - b^2$, indicating that $a > b$. This is important in that it helps us recognize the orientation of the ellipse from the equation.

Graph	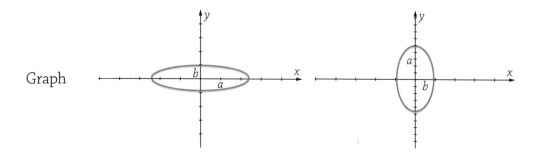	
Equation	$\dfrac{x^2}{a^2} + \dfrac{y^2}{b^2} = 1$	$\dfrac{x^2}{b^2} + \dfrac{y^2}{a^2} = 1$

The value $2a$ represents the length of the major axis, while the value $2b$ represents the length of the minor axis. The foci are always on the major axis.

▶ Determine the equation of the ellipse with foci at (–3, 0) and (3, 0) and whose major axis has length 10.

▶ The center of the ellipse is at the origin, $c = 3$, and $a = 5$. Therefore, $b = 4$ (since $b^2 = a^2 - c^2$).

▶ The equation of the ellipse is $\dfrac{x^2}{25} + \dfrac{y^2}{16} = 1$.

Here is another example

▶ Determine the equation of the ellipse with foci at (–3, 4) and (–3, 20) and whose minor axis has a length of 4.

▶ The major axis is vertical in this example. The center of the ellipse is at (–3, 12), so the value of $c = 8$. Using $b^2 = a^2 - c^2$, we get $16 = a^2 - 64$, so $a^2 = 80$.

▶ The equation of the ellipse is $\dfrac{x^2}{16} + \dfrac{y^2}{80} = 1$.

The ellipse also has a focal chord (in fact there are two *focal chords*). The focal chords are the segments perpendicular to the major axis with its endpoints on the ellipse. The length of each focal chord is $\dfrac{2b^2}{a}$.

▶ Determine the coordinates of the focal chords for the ellipse $\dfrac{x^2}{25} + \dfrac{y^2}{16} = 1$.

▶ We know the foci are located at the points $(\pm 3, 0)$. Substituting $x = 3$ into the equation, we get $\dfrac{9}{25} + \dfrac{y^2}{16} = 1$, which becomes

$\dfrac{y^2}{16} = \dfrac{16}{25}$. Therefore, $y^2 = \dfrac{16^2}{25} \Rightarrow y = \pm\dfrac{16}{5}$.

▶ The coordinates of the endpoints of the focal chords are

$\left(3, \dfrac{16}{5}\right), \left(3, \dfrac{-16}{5}\right), \left(-3, \dfrac{16}{5}\right)$, and $\left(-3, \dfrac{-16}{5}\right)$.

This is an example of working with an equation of the parabola in standard form and rewriting it in the more convenient vertex form:

▶ Given the ellipse with the equation $144x^2 + 169y^2 + 864x - 676y - 22{,}364 = 0$, give a complete analysis, including the coordinates of the center, foci, vertices, endpoints of the major axis, endpoints of the minor axis, and endpoints of the focal chords.

▶ Gather terms in x together, gather terms in y, and move the constant to the right.

$$144x^2 + 864x + 169y^2 - 676y = 22{,}364$$

▶ Complete the square in x and y.

$$144(x^2 + 6x) + 169(y^2 - 4y) = 22{,}364$$

$$144(x^2 + 6x + 9) + 169(y^2 - 4y + 4) = 22{,}364 + 144(9) + 169(4)$$

$$144(x + 3)^2 + 169(y - 2)^2 = 24{,}336$$

$$\frac{144(x + 3)^2}{24{,}336} + \frac{169(y - 2)^2}{24{,}336} = 1$$

$$\frac{(x + 3)^2}{169} + \frac{(y - 2)^2}{144} = 1$$

▶ The coordinates for the center of the ellipse are $(-3, 2)$. The major axis of the ellipse is horizontal (because the larger denominator is under the term in x) with $a = 13$. The endpoints of the major axis are $(-16, 2)$

and (10, 2). With $b = 12$, the endpoints of the minor axis are (–3, –10) and (–3, 14).

▶ The distance from the center of the ellipse to each focus is 5 units. The coordinates of the foci are (–8, 2) and (2, 2). The distance from a focus to the end of a focal chord is $\dfrac{b^2}{a}$. In this example, that will be $\dfrac{144}{13}$.

▶ The endpoints for the focal chords are $\left(2, \dfrac{170}{13}\right)$, $\left(2, \dfrac{-118}{13}\right)$, $\left(-8, \dfrac{170}{13}\right)$, and $\left(-8, \dfrac{-118}{13}\right)$.

I had the good fortune to be able to tour the Colosseum when vacationing in Rome. It was incredible to me to see how the gates to the Colosseum were numbered (Roman numerals, of course!) and that each gate brought you to a specific seating section, very much like the gates used in arenas today.

EXAMPLE

▶ The Roman Colosseum is an ellipse with a major axis of 620 feet and a minor axis of 513 feet. Find the distance between the foci of this ellipse to the nearest tenth.

Photo by the author

▸ The dimensions tell us that $2a = 620$ and $2b = 513$, so $a = 310$ and $b = 256.5$. Use $c^2 = a^2 - b^2$ to determine that $c = 174.1$.

▸ The foci of the structure are 348.2 feet apart.

The ellipse has an interesting reflective property. Any ray (sound or light) that passes through one focus will also pass through the other. A ray that does not pass through a focus never will rebound to pass through either focus.

IRL   There are lots of ellipses around us—we're even part of one or two!

The Ontario Science Centre, in Toronto, Canada, used to have an area set up as an ellipse. No matter what noise was occurring within (or outside) this space, if two people were standing at the clearly marked foci, they could hear each other. The United States Capitol Building and St. Paul's Cathedral in London each have elliptical domes and have whispering spaces at the foci. There is even a story that John Quincy Adams used to pretend to sleep at one of the foci in the Capitol Building while his opponents conferred at the other focus.

You may have played on an elliptical pool table. On an elliptical pool table, there is only one pocket, located at one of the foci. The second focus is clearly marked. As you might guess, one merely needs to be sure the ball being struck crosses over the foci and rebounds off the bumper to go into the pocket. The only obstacle will be another ball in the path.

While not a reflective property, another application of the ellipse is planetary motion. The path each planet takes as it travels around the Sun is that of an ellipse. The Sun sits at one of the foci of Earth's orbit.

Analyze and Graph Hyperbola

When a plane cuts both cones and does not contain the common vertex, the cross section is a **hyperbola**.

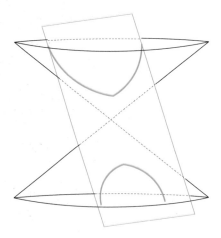

The analytic definition of an ellipse is that it is the locus of points from two fixed points so that the absolute value of the differences of the distances is a constant. The derivation for the equation of the hyperbola is similar to that of the ellipse.

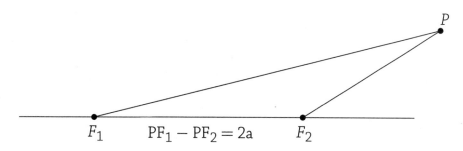

Once again, we will designate the coordinates for the foci to be $(-c, 0)$ and $(c, 0)$. The locus definition generates the equation $\sqrt{(x + c)^2 + y^2} - \sqrt{(x - c)^2 + y^2} = 2a$. Repeat the steps to remove the radicals from the equation.

$$\sqrt{(x + c)^2 + y^2} = 2a + \sqrt{(x - c)^2 + y^2}$$

$$\left(\sqrt{(x + c)^2 + y^2}\right)^2 = \left(2a + \sqrt{(x - c)^2 + y^2}\right)^2$$

$$(x + c)^2 + y^2 = 4a^2 + 4a\sqrt{(x - c)^2 + y^2} + (x - c)^2 + y^2$$

$$x^2 + 2cx + c^2 + y^2 = 4a^2 + 4a\sqrt{(x - c)^2 + y^2} + x^2 - 2cx + c^2 + y^2$$

$$4cx - 4a^2 = 4a\sqrt{(x - c)^2 + y^2}$$

$$cx - a^2 = a\sqrt{(x - c)^2 + y^2}$$

$$(cx - a^2)^2 = \left(a\sqrt{(x - c)^2 + y^2}\right)^2 = a^2\left((x - c)^2 + y^2\right)$$

$$c^2x^2 - 2a^2cx + a^4 = a^2x^2 - 2a^2cx + a^2c^2 + a^2y^2$$

$$c^2x^2 - a^2x^2 - a^2y^2 = a^2c^2 - a^4$$

$$(c^2 - a^2)x^2 - a^2y^2 = a^2(c^2 - a^2)$$

Let $c^2 - a^2 = b^2$: $b^2x^2 - a^2y^2 = a^2b^2$

$$\frac{x^2}{a^2} - \frac{y^2}{b^2} = 1$$

The center of the hyperbola is the origin and the x-intercepts are at $(\pm a, 0)$. (Observe that $c > a$ for the hyperbola.) How does the value of b fit into the equation? As is often the case, a picture helps explain it.

The graph of the hyperbola $\dfrac{x^2}{16} - \dfrac{y^2}{9} = 1$ is shown.

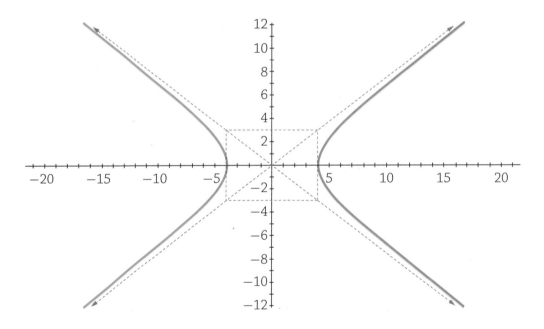

The points $(-4,0)$ and $(4, 0)$ are the x-intercepts but also the points where the hyperbola intersects the line containing the foci. These points are the **vertices** of the hyperbola. A rectangle is formed from the vertices with dimensions $2a$ by $2b$. The diagonals of this rectangle form the oblique asymptotes for the hyperbola. The segment joining the two vertices is called the **transverse axis**, while the segment from $(0, -b)$ to $(0, b)$ is called the **conjugate axis**.

Before we look at a few examples, we need to go back to the asymptotes. The oblique asymptote describes the end behavior of the function. We rewrite the equation of the hyperbola as $\dfrac{y^2}{b^2} = \dfrac{x^2}{a^2} - 1 \Rightarrow y^2 = \dfrac{b^2 x^2}{a^2} - b^2$.

As $x \to \pm\infty$, $y^2 \to \dfrac{b^2 x^2}{a^2}$ so that $y \to \pm\dfrac{bx}{a}$.

Graph	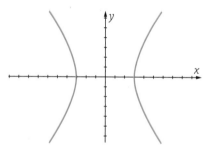	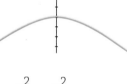
Equation	$\dfrac{x^2}{a^2} - \dfrac{y^2}{b^2} = 1$	$\dfrac{y^2}{a^2} - \dfrac{x^2}{b^2} = 1$
Foci	$(\pm c, 0)$	$(0, \pm c)$
Vertices	$(\pm a, 0)$	$(0, \pm a)$
Asymptotes	$y = \pm\dfrac{bx}{a}$	$y = \pm\dfrac{ax}{b}$

▶ Determine the equation of the hyperbola foci having coordinates (–3, 2) and (7, 2) and the length of the transverse axis is 8. In addition, determine the equations of the asymptotes for the hyperbola.

▶ The transverse axis is horizontal, so the equation is of the form $\dfrac{x^2}{a^2} - \dfrac{y^2}{b^2} = 1$. The center of the hyperbola is (2, 2) and the value is $c = 5$. The length of the transverse axis is 8, so $a = 4$. Using $b^2 = c^2 - a^2$ tells us that $b = 3$.

▶ Therefore, the equation of the hyperbola is $\dfrac{(x-2)^2}{16} - \dfrac{(y-2)^2}{9} = 1$.

The equations for the asymptotes are $y - 2 = \pm\dfrac{3}{4}(x - 2)$.

▶ Remember, as $x \to \pm\infty$, $(y-2)^2 \to \dfrac{9(x-2)^2}{16}$.

The next example asks us to take an equation in standard form and rewrite it in the vertex form (similar to the example in the section on ellipses). A word of warning: *be careful of the negative*. Even though we are a long way from the Algebra I days, not paying attention to the negative was an error students made when I was teaching that resulted in wrong answers.

▶ Given the hyperbola with equation $16x^2 - 9y^2 + 64x + 108y - 116 = 0$, determine the coordinate of the center of the hyperbola, the coordinates of the foci and vertices, the coordinates of the endpoints of the focal chords, and the equations of the asymptotes.

▶ Gather terms in x together, gather terms in y, and move the constant to the right.

$$16x^2 - 9y^2 + 64x + 108y = 116$$

▶ Complete the square:

$$16(x^2 + 4x) - 9(y^2 - 12y) = 116$$

$$16(x^2 + 4x + 4) - 9(y^2 - 12y + 36) = 116 + 64 - 324$$

$$16(x + 2)^2 - 9(y - 6)^2 = -144$$

$$\frac{16(x + 2)^2}{-144} - \frac{9(y - 6)^2}{-144} = 1$$

$$\frac{(y - 6)^2}{16} - \frac{(x + 2)^2}{9} = 1$$

▶ The center of the hyperbola is (–2, 6). Since $a^2 = 16 \Rightarrow a = 4$ and $b^2 = 9 \Rightarrow b = 3$, we can determine that $c = 5$.

▶ The term in y is positive, so the transverse axis is vertical. Therefore, the vertices of the hyperbola have coordinates (–2, 10) and (–2, 2) and the foci have coordinates (–2, 11) and (–2, 1). To determine the coordinates for the endpoints of the focal chords, first set $y = 1$ and solve for x.

$$\frac{(1 - 6)^2}{16} - \frac{(x + 2)^2}{9} = 1 \Rightarrow \frac{25}{16} - \frac{(x + 2)^2}{9} = 1$$

$$\Rightarrow \frac{25}{16} - 1 = \frac{(x + 2)^2}{9} \Rightarrow \frac{9}{16} = \frac{(x + 2)^2}{9} \Rightarrow \frac{81}{16} = (x + 2)^2$$

$$\Rightarrow x + 2 = \pm\frac{9}{4}.$$

▶ The endpoints of the focal chords will be $\dfrac{9}{4}$ to the left and right of the foci, making the endpoints of the focal chords $\left(\dfrac{-17}{4},1\right)$, $\left(\dfrac{1}{4},1\right)$, $\left(\dfrac{-17}{4},11\right)$, $\left(\dfrac{1}{4},11\right)$. (We can include the endpoints for the focal chord through (–2, 11) because $(1-6)^2 = (11-6)^2$ so we'll get the same solution for x.) We can also deduce that the length of the focal chord is $\dfrac{2b^2}{a}$.

▶ To determine the equations of the asymptotes, I suggest that rather than memorizing the different sets of formulas, you realize that you are examining end behavior:

$$\frac{(y-6)^2}{16} - \frac{(x+2)^2}{9} = 1 \Rightarrow \frac{(y-6)^2}{16} - 1 = \frac{(x+2)^2}{9}$$

$$\Rightarrow \frac{9(y-6)^2}{16} - 9 = (x+2)^2 \Rightarrow x+2 = \pm\sqrt{\frac{9(y-6)^2}{16} - 9}$$

▶ As the values of x get very large, the -9 becomes negligible and the asymptotes are $x+2 = \pm\dfrac{3}{4}(y-6)$.

This next problem has partial information from which you must deduce what you will need to finish the problem.

EXAMPLE

▶ Determine the equation of the hyperbola with a vertex at (6, 5), the conjugate axis is parallel to the x-axis, and whose asymptotes have equations $5x - 6y = 30$ and $5x + 6y = 30$.

▶ The asymptotes always intersect at the center of the hyperbola. Solving the system of equation tells us the center is at the point (6, 0). This tells us that $a = 5$ and the second vertex has coordinates $(6, -5)$. (This agrees with the statement that the conjugate axis is horizontal, making the transverse axis vertical.) The slopes of the asymptotes are $\pm\dfrac{5}{6}$. We can conclude that $b = 6$.

▶ Therefore, the equation of the hyperbola is $\dfrac{y^2}{25} - \dfrac{(x-6)^2}{36} = 1$.

The hyperbola also has an interesting reflective property. When a ray passes through a focus of the hyperbola and then reflects off the hyperbola, the ray of reflection appears to come from the other focus.

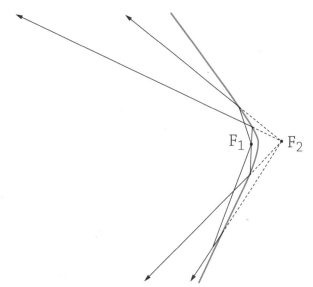

This property is applied in brass instruments and in speakers. The sounds are dispersed over a wide range, unlike the parabola, which concentrates the rays into a tight area.

BTW

An interesting video on YouTube called "Stand Up Conic" (!) from the series "For All Practical Purposes" does a very good job of showing how the conic sections are used.

At this point, we have made sure that the directrix for the parabola, the major and minor axes for the ellipse, and the transverse and conjugate axes for the hyperbola have either been horizontal or vertical. The only thing that keeps us limited to these conditions is algebra. Rotating these lines requires a bit of trigonometry, which we'll look at later in this book. That is, we'll look at these later, but with one exception. If k is a constant, the graph of the equation $xy = k$ is a hyperbola whose asymptotes are the x and y axes. If $k > 0$, the graph will lie in the first and third quadrants, while if $k < 0$, the graph will lie in the second and fourth quadrants.

Eccentricity

The **eccentricity** of a conic section helps describe the curve. Eccentricity is defined as the ratio of the distance of a point on the curve to a fixed point (focus) and a fixed line (directrix). For a matter of interpretation, the eccentricity is an indication of how much the curve deviates from a circle. The eccentricity of a circle is 0 because the circle does not deviate from itself. For the remaining conic sections, the formula $e = \dfrac{c}{a}$ is used to compute the eccentricity. For the parabola, $e = 1$ (points are equidistant from the focus and directrix), while the eccentricity of the ellipse is between 0 and 1 (that is, $0 < e < 1$) and the eccentricity of the hyperbola is greater than 1. This should make sense. The focus is closer to the center of the ellipse than is the vertex, and the reverse is true for the hyperbola.

EXAMPLE

▶ Determine the equation of the conic with center at (2, 2), vertex at (8, 2), and eccentricity $\dfrac{2}{3}$.

▶ Since $0 < e < 1$, the conic is an ellipse. Given the coordinates of the center and vertex, we know the major axis is horizontal. The distance from the center to the focus is $a = 6$. Given this, we get $\dfrac{2}{3} = \dfrac{c}{6}$; so it follows that $c = 4$. We can now determine the value of b^2 is 20.

▶ The equation of the ellipse is $\dfrac{(x-2)^2}{36} + \dfrac{(y-2)^2}{20} = 1$.

This second example involving the eccentricity gives us only the foci.

EXAMPLE

▶ Determine the equation of the conic with foci at $(3, -3)$ and $(3, 7)$ and has eccentricity $\dfrac{5}{3}$.

▶ The eccentricity is greater than 1, so the conic is a hyperbola. The foci lie on the line $x = 3$, so we know the transverse axis is vertical. The distance between the two foci is 10. Therefore, $c = 5$. Use the formula for eccentricity to show that $a = 3$. Consequently, $b = 4$.

▶ The equation for the conic is $\dfrac{(y-2)^2}{9} - \dfrac{(x-3)^2}{16} = 1$.

Eccentricity will come back into play when we discuss the polar form for the equations of the conics in Chapter 10.

Solutions of Systems of Quadratic Equations

At this point in your mathematical career, you have solved systems of equations involving two (or more) linear equations, a linear and quadratic equation, and, possibly, a circle and a parabola. We will now examine systems of quadratic equations involving all the conic sections.

▶ Solve for x and y: $x^2 + y^2 = 16$

$$2x^2 - 3y^2 = 12$$

▶ We solve this using the elimination method.

$$3(x^2 + y^2 = 16) \Rightarrow 3x^2 + 3y^2 = 48$$
$$2x^2 - 3y^2 = 12 \qquad 2x^2 - 3y^2 = 12$$

▶ Add to get $5x^2 = 60$ so $x^2 = 12$ and $x = \pm 2\sqrt{3}$.

▶ Solve for y: $x^2 + y^2 = 16$ becomes $12 + y^2 = 16$ and $y = \pm 2$.

▶ Therefore, the solution is $(-2\sqrt{3}, -2), (-2\sqrt{3}, 2), (2\sqrt{3}, -2)$, and $(2\sqrt{3}, 2)$.

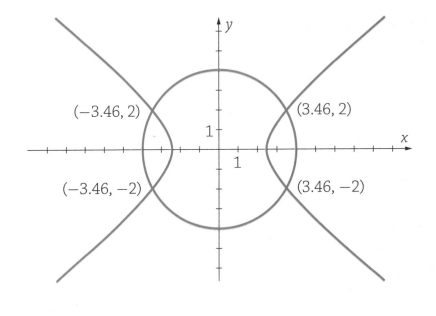

An example of an ellipse intersecting a circle is next.

EXAMPLE

▶ Solve for x and y: $x^2 + y^2 = 10$

$$12x^2 + 9y^2 = 108$$

▶ Solve by elimination: $\begin{aligned} -9(x^2 + y^2 = 10) \Rightarrow -9x^2 - 9y^2 = -90 \\ 12x^2 + 9y^2 = 108 \quad 12x^2 + 9y^2 = 108 \end{aligned}$

▶ Add to get $3x^2 = 18$ so $x^2 = 6$ and $x = \pm\sqrt{6}$.

▶ Solve for y: $x^2 + y^2 = 10$ becomes $6 + y^2 = 10$, $y^2 = 4$, and $y = \pm 2$.

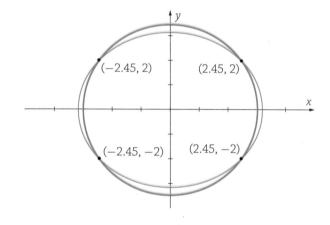

This last question is a bit more involved.

EXAMPLE

▶ The line $3x + 4y = 50$ is tangent to a circle with radius 5 at the point $(6, 8)$. Determine the equations for the circles that satisfy these conditions.

▶ The language of the question tells us that there is more than one circle that satisfies the problem. The slope of the tangent line is $\dfrac{-3}{4}$, so the line containing the center of the circle(s) and the point of tangency must be $\frac{4}{3}$. The equation of the line with slope $\frac{4}{3}$ passing through the point (6, 8) is $y - 8 = \frac{4}{3}(x - 6)$ or $y = \frac{4}{3}x$. If the coordinates for the center of the circle are (h, k), then it must be true that $k = \frac{4}{3}h$ and $(6 - h)^2 + (8 - k)^2 = 25$. Substitute for k to get

$(6 - h)^2 + \left(8 - \frac{4h}{3}\right)^2 = 25$. Solve this quadratic equation.

$$(6 - h)^2 + \left(8 - \frac{4h}{3}\right)^2 = 25$$

$$(h - 6)^2 + \left(\frac{4}{3}(h - 6)\right)^2 = 25$$

$$(h - 6)^2 + \frac{16}{9}(h - 6)^2 = 25$$

$$\frac{25}{9}(h - 6)^2 = 25$$

$$(h - 6)^2 = 9$$

$$h - 6 = \pm 3$$

$$h = 3, 9$$

▶ If $h = 3$, then $k = 4$, and the equation of the circle would be $(x - 3)^2 + (y - 4)^2 = 25$.

▶ If $h = 9$, then $k = 12$, and the equation of the circle would be
$(x-9)^2 + (y-12)^2 = 25$.

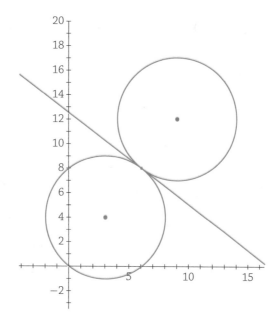

EXERCISES

EXERCISE 5-1

Determine the equation of the circle described in each question.

1. Determine the equation of the circle that has a center at $(5, 6)$ and is tangent to the x-axis.

2. Determine the equation of the circle that has a diameter with endpoints at $(-4, 7)$ and $(12, 15)$.

3. Determine the coordinates of the intercepts of the circle with equation $(x + 2)^2 + (y - 5)^2 = 40$.

EXERCISE 5-2

Solve these problems involving parabolas.

1. Determine the equation of the parabola with focus at $(4, 6)$ and directrix at $y = 2$.

2. Determine the equation of the parabola with focus at $(4, 6)$ and directrix at $x = 2$.

3. Determine the equation of the parabola with focus at $(-5, 2)$ and vertex at $(-3, 2)$.

4. Determine the equation of the directrix and coordinates of the focus for the parabola with equation $y + 6 = \dfrac{-1}{20}(x - 3)^2$.

5. Determine the equation of the directrix and coordinates of the focus for the parabola with equation $x - 1 = \dfrac{1}{12}(y - 4)^2$.

EXERCISE 5-3

Determine the equation of the ellipse described.

1. Foci at (–3, 5) and (1, 5) and major axis has length 10.

2. Endpoints of the major and minor axes at (2, 12), (–3, 4), (2, –4), (7, 4).

3. Center at (5, 6), a focus at (10, 6), and the endpoint of a major axis at (13, 6).

EXERCISE 5-4

Determine the coordinates of the center, the foci, the endpoints of the major and minor axes, and the eccentricity.

1. $x^2 + 4y^2 + 2x - 8y + 1 = 0$

2. $9x^2 + 4y^2 + 36x - 8y + 4 = 0$

3. $36x^2 + 100y^2 + 360x - 800y - 1100 = 0$

4. A one-way road passes under an overpass in the form of half of an ellipse, 15 feet high at the center and 20 feet wide. Assuming a truck is 12 feet wide, what is the height of the tallest truck that can pass under the overpass without getting stuck?

EXERCISE 5-5

Use this information to answer the following questions:

   Statuary Hall is an elliptical room in the United States Capitol in Washington, D.C. The room is also called the Whispering Gallery because a person standing at one focus of the room can hear even a whisper spoken by a person standing at the other focus. This occurs because any sound that is emitted from one focus of an ellipse will reflect off the side of the ellipse to the other focus. Statuary Hall is 46 feet wide and 97 feet long.

1. , Find an equation that models the shape of the room.

2. How far apart are the two foci?

3. How wide is the room at each focus?

EXERCISE 5-6

Determine the equation of the hyperbola described.

1. Vertices at (1, 2) and (1, –2) and the length of the conjugate axis is 6.

2. Center at (–2, 2), a focus at (6, 2), and a vertex at (4, 2).

3. Center at (4, 2), a vertex at (4, 6), and one asymptote with equation $4x - 3y = 10$.

EXERCISE 5-7

Determine the coordinates of the center, the vertices, the foci, and the endpoints of the focal chords, as well as the equations of the asymptotes and the eccentricity.

1. $x^2 - 2y^2 + 6x + 4y + 5 = 0$

2. $4x^2 - 49y^2 + 48x - 98y + 291 = 0$

3. $9x^2 - 4y^2 + 36x - 16y - 16 = 0$

EXERCISE 5-8

Solve the system of equations in each question.

1. $y = x^2 - 1$

$2x^2 + y^2 = 2$

2. $3y^2 - 4x^2 = 12$

   $16x^2 + 9y^2 = 144$

3. $x^2 + y^2 = 25$

   $xy = 12$

One more!

4. The line $12x + 5y = 338$ is tangent to two circles of radius 13 at the point (24, 10). Determine the equations of both circles.

Flashcard App

Exponential and Logarithmic Functions

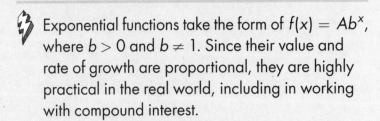

Exponential functions take the form of $f(x) = Ab^x$, where $b > 0$ and $b \neq 1$. Since their value and rate of growth are proportional, they are highly practical in the real world, including in working with compound interest.

Logarithms help us solve exponential functions. Logarithmic scales find practical use in measuring phenomena such as earthquakes and sound.

lbert Einstein is credited with saying that "compound interest is the most powerful force in the universe." Why is this? The entire financial world is based on compound interest. And what is compound interest? It is an exponential function.

Exponential functions and their inverses, logarithmic functions, are necessary tools to describe complex events. In this chapter, we'll look at the basic properties of these functions and some applications that are used or are in the news almost daily.

Exponential Functions

We rely on the basic properties of exponents when we work with exponential functions. A summary of the properties is shown in the following table.

Rule	Description	Example
$(b^m)(b^n) = b^{m+n}$	When multiplying terms with a common base, add the exponents.	$(5^7)(5^4) = 5^{11}$
$(ab)^m = a^m b^m$	A product raised to a power is the same as the product of the result when each base is raised to the power.	$(5x)^4 = 5^4 x^4$
$(a^n)(b^n) = (ab)^n$	When multiplying terms with a common exponent, multiply the bases.	$(3^7)(2^7) = 6^7$
$\dfrac{b^m}{b^n} = b^{m-n}$	When dividing terms with a common base, subtract the exponents.	$\dfrac{12^9}{12^4} = 12^5$
$b^0 = 1; (b \neq 0)$	One of two immediate consequences of the quotient rule, any nonzero number raised to the zero power is equal to 1.	$12000000^0 = 1$
$b^{-n} = \dfrac{1}{b^n}$	A second consequence of the quotient rule is that a term raised to a negative exponent is the reciprocal of the term to the positive power.	$\dfrac{12^4}{12^9} = \dfrac{1}{12^5} = 12^{-5}$

$(b^m)^n = b^{mn}$	This is a consequence of the product rule. When a term raised to a power is then raised to another power, the result is equal to the base raised to the product of the powers.	$(5^7)^4 = 5^{28}$
$b^{\frac{1}{n}} = \sqrt[n]{b}$	Fractional exponents represent radical expressions.	$5^{\frac{1}{3}} = \sqrt[3]{5}$

Let's check out an example.

EXAMPLE

▶ Simplify $\left(\dfrac{1000x^{-4}y^{-8}}{27x^{-10}y^4}\right)^{\frac{-2}{3}}$.

▶ As is usually the case, first work with what is inside the parentheses.

$$\left(\frac{1000x^{-4}y^{-8}}{27x^{-10}y^4}\right)^{\frac{-2}{3}} = \left(\frac{1000x^6}{27y^{12}}\right)^{\frac{-2}{3}}$$

▶ The negative exponent outside the parentheses indicates a reciprocal is involved.

$$\left(\frac{1000x^6}{27y^{12}}\right)^{\frac{-2}{3}} = \left(\frac{27y^{12}}{1000x^6}\right)^{\frac{2}{3}}$$

▶ Separate the fractional exponents.

$$\left(\frac{27y^{12}}{1000x^6}\right)^{\frac{2}{3}} = \left(\left(\frac{27y^{12}}{1000x^6}\right)^{\frac{1}{3}}\right)^2$$

▶ Fractional exponents mean roots.

$$\left(\left(\frac{27y^{12}}{1000x^6}\right)^{\frac{1}{3}}\right)^2 = \left(\frac{3y^4}{10x^2}\right)^2$$

▶ Finish the problem.

$$\left(\frac{1000x^{-4}y^{-8}}{27x^{-10}y^4}\right)^{\frac{-2}{3}} = \left(\frac{3y^4}{10x^2}\right)^2 = \frac{9y^8}{100x^4}$$

Functions of the form $f(x) = Ab^x$, with $b > 0$ and $b \neq 1$, are called **exponential functions**. (The value of b is not allowed to be 1 because 1 raised to any power is 1 and the function would be a constant.) The graphs of exponential functions have the characteristics that the y-intercept is always $(0, A)$ and it passes through the point $(1, Ab)$, the domain is the set of real numbers, the function is $1 = 1$, and the graph has a horizontal asymptote at $y = 0$. If $A > 0$, the range of the function is $y > 0$ (the range is $y < 0$ if $A < 0$). When $b > 1$, the graph increases in value from left to right, and if $0 < b < 1$, the graph decreases in value.

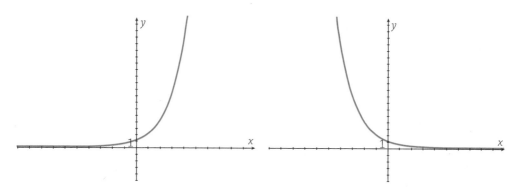

While any positive number can serve as the base of an exponential function (we already excluded 1, remember), there are three numbers that are used more than the others. Ten, because it is the base of our number system, is one of the three. Two, because many decay problems refer to half-life, is the second. The third is tied to the story of compound interest.

Compound interest is a relatively new item in the story of money lending (which goes back for millennia). Leonhard Euler, a Swiss mathematician, while investigating the process of computing compound interest asked himself whether there was a maximum amount that one could be charged at a given rate of interest. That is to say, he assumed that the rate of interest was 100%, so the formula for computing the compound interest for 1 monetary unit was $\left(1 + \dfrac{1}{n}\right)^n$ with the number of compound periods in one year being n. Looking at various values for n, he saw that there was a limiting value.

n	Value
2	2.25
4	2.44141
12	2.61304
24	2.66373
100	2.70481
1000	2.71692
10,000	2.71815
1,000,000	2.71828

No matter how large the value of n became, the value of the compound interest stayed at approximately 2.71828. This number turned out to be as important in the world of mathematics as the number π. In honor of the work Euler did, this number is called **Euler's number**, or e. While almost everyone can tell you that π is approximately 3.14 (and some can say

3.14159—or more), it is worth your while to know that e is approximately 2.718.

If the number e is not one that you have used much, take a moment to use your calculator to get a feel for its magnitude. (If you keep in mind that e is between 2 and 3, then you will not be too surprised at the results.)

EXAMPLE

▶ For what integer values of n will $e^n > 1000$?

▶ This is a trial-and-error problem at this point. Do you know that a kilobyte is really not 1000 bytes but is 1024 bytes because $2^{10} = 1024$?

▶ By entering e^{10}, e^8, e^7, you see that when $n > 7$, $e^n > 1000$.

Solving exponential equations can be done in a number of ways. The most basic method is to rewrite the equations so that they have the same base.

EXAMPLE

▶ Solve: $2^x = 32$.

▶ Since 32 is a power of 2, we rewrite the problem to be $2^x = 2^5$, so $x = 5$.

The next example is a similar problem, but one that requires an extra step.

EXAMPLE

▶ Solve $8^{2x+3} = 32^{x+4}$.

▶ Thirty-two is not a convenient power of eight, but both 32 and 8 are convenient powers of 2. Rewrite the example with the common base: $8^{2x+3} = 32^{x+4} \Rightarrow (2^3)^{2x+3} = (2^5)^{x+4}$

▶ Manipulate the exponents: $2^{6x+9} = 2^{5x+20}$

▶ Set the exponents equal: $6x + 9 = 5x + 20 \Rightarrow x = 11$

With e being a **transcendental number** (a number that is not a root of a polynomial equation with real coefficients), it is impossible to solve an exponential equation using e unless it is of the form $e^a = e^b$ if the expectation is that the equation will use algebraic means. However, such equations are easily solved if a graphic solution is considered.

EXAMPLE

▶ Solve $e^{x-2} + 2 = 4^{x+3}$.

▶ The resulting graphic solution is as follows:

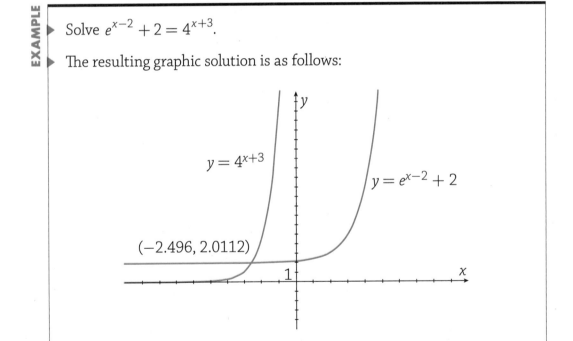

As a last example for equations involving exponential functions, this has a bit of a twist to it.

EXAMPLE

▶ Solve $9^{2x} - 30(9^x) + 81 = 0$.

▶ Rather than getting nervous about the seemingly complex nature of the problem, take a moment to look at it. There are three terms and there is a variable expression in the middle term. How does the first term compare to this middle term? Yes, indeed, it is the square of the algebraic expression in the middle term. This problem is really no different than any other trinomial. If you wish, substitute a for the exponential expression.

▶ The problem now becomes: $a^2 - 30a + 81 = 0$, which you know will factor to be $(a - 3)(a - 27) = 0$. Working with the exponential expression again, we need to solve the equations $9^x = 3$ and $9^x = 27$. Writing the problems as powers of 3, $3^{2x} = 3$ and $3^{3x} = 3^3$.

▶ Solving these equations, we get $x = \dfrac{1}{2}, 1$.

Logarithmic Functions

All exponential functions are $1 - 1$ and consequently have inverse functions. Given the function $f: y = b^x$, the inverse function is $f^{-1}: x = b^y$. However, unlike the previous examples of inverse functions that we have solved, there is no algebraic manipulation that we have seen that allows us to solve for y in this equation. It is by definition that we get this new mathematical function called the **logarithm** and, by this definition, $x = b^y \Leftrightarrow y = \log_b(x)$. You

BTW

A Scottish preacher and astronomer, John Napier, was the first to name this new function. He called them **logarithms**, *derived from the Greek words for ratio and number.*

should be able to see statements such as $\log_2(32)$–which is read as the base 2 log of 32–and know that it means "to what power must 2 be raised in order to get 32?"

With the logarithmic function being the inverse of the exponential function, a lot of the properties should be the same. These are some of the most critical properties:

Rule	Description
$\log_b(mn) = \log_b(m) + \log_b(n)$	The exponent of a product is the sum of the exponents.
$\log_b\left(\dfrac{m}{n}\right) = \log_b(m) - \log_b(n)$	The exponent of a quotient is the difference of the exponents.
$\log_b(m^n) = n\log_b(m)$	When a term raised to a power is then raised to another power, the exponent is the product of the exponents.
$\log_b(1) = 0$	If (0, 1) is a point on the exponential function, then (1, 0) is a point of the logarithmic function.
$\log_b(b) = 1$	b raised to what power is b? It has to be 1.

Let's see how these rules apply.

▶ Assume $\log_b(2) = x$, $\log_b(3) = y$, and $\log_b(5) = z$. Express $\log_b(90b^2)$ in terms of x, y, and z.

▶ We know that $90 = (9)(10) = (3^2)(2)(5)$. Use this to write: $\log_b(90b^2)$ as $\log_b(2 \times 3^2 \times 5b^2)$.

▶ Apply the properties of logarithms: $\log_b(2 \times 3^2 \times 5b^2) = \log_2(2) + \log_2(3^2) + \log_2(5) + \log_2(b^2) = \log_2(2) + 2\log_2(3) + \log_2(5) + 2\log_2(b) = x + 2y + z + 2$.

This next example is not much more difficult, but there is a part where you must be careful.

Express $\log_b\left(\dfrac{x^3\sqrt{y}}{b^5z^4}\right)$ in terms of $\log_b(x), \log_b(y),$ and $\log_b(z)$.

We will apply the quotient rule first, $\log_b\left(\dfrac{x^3\sqrt{y}}{b^5z^4}\right) = \log_b(x^3y^{\frac{1}{2}}) - $

$\log_b(b^5z^4)$, noting that the radical was written as a fractional exponent:

$\log_b(x^3y^{\frac{1}{2}}) - \log_b(b^5z^4) = \log_b(x^3) + \log_b(y^{\frac{1}{2}}) - \left(\log_b(b^5) + \log_b(z^4)\right) \cdot$
$\log_b(x^3) + \log_b(y^{\frac{1}{2}}) - \left(\log_b(b^5) + \log_b(z^4)\right) = 3\log_b(x) + \dfrac{1}{2}\log_b(y) - $
$5 - 4\log_b(z) \cdot$

Too often the term in z will have a plus sign in front of it because people were not taking the time to write the parentheses around the product that is in the denominator of the fraction.

This next example illustrates an important property of logarithms.

Assume $\log_b(2) = x$ and $\log_b(3) = y$. Express $\log_2(3)$ in terms of x and y.

We'll assign the variable q to be the value of $\log_2(3)$. If $\log_2(3) = q$, the definition of the logarithm tells us that $2^q = 3$. Apply base b logarithms to both sides of the equation: $\log_b(2^q) = \log_b(3) \Rightarrow$ $q\log_b(2) = \log_b(3)$.

Substituting x and y, the equation becomes $qx = y$ so that $q = \dfrac{y}{x}$.

The result of this example, $\log_2(3) = \dfrac{\log_b(3)}{\log_b(32)} = \dfrac{y}{x}$, is an example of the change of base formula. In general, the **change of base formula** says $\log_m(n) = \dfrac{\log_b(n)}{\log_b(m)}$.

 IRL In days of yore, like the year 2000, there were two logarithm buttons on a calculator. There were common logarithms (base 10 logarithms) and natural logarithms (base e logarithms). Common logs were abbreviated log (notice that the base is assumed), while natural logs were written as *ln*. (Think Romance languages in which the noun is followed by the adjective, as in *Moulin Rouge*.) The calculators currently on the market are designed to accommodate any base when you press the log button. Consequently, whenever you have to determine the decimal value of a logarithm, no matter the base, you can use your calculator to do so. The change of base formula is now mostly used with algebraic manipulation.

Now let's look at solving equations with logarithms.

EXAMPLE

▶ Solve $\log_4(x + 3) = 2$.

▶ Always remember: a logarithm is an exponent. When you are solving logarithmic equations, the goal is to convert the problem to an exponential equation that you can manipulate.

▶ Given this statement: $\log_4(x + 3) = 2 \Rightarrow x + 3 = 4^2 = 16$. Therefore, $x = 13$.

That wasn't too bad. Here is another fairly straightforward problem.

EXAMPLE

▶ Solve $\log_5(2x-3)=3$.

▶ Write as an exponential equation: $2x-3=5^3=125$

$$2x=128 \Rightarrow x=64$$

The next few problems require that we use the properties of logarithms before we are able to create the exponential equation.

EXAMPLE

▶ Solve $\log_6(x+4)+\log_6(x-5)=2$.

▶ We need to combine the terms on the left side of the equation into a single logarithm.

$$\log_6(x+4)+\log_6(x-5)=\log_6\big((x+4)(x-5)\big)$$

$$\log_6\big((x+4)(x-5)\big)=2 \Rightarrow x^2-x-20=6^2$$

$$x^2-x-20=6^2 \Rightarrow x^2-x-56=0$$

$$(x-8)(x+7)=0 \Rightarrow x=8,-7$$

▶ We've been a little slack about checking the solutions to the problems we have done so far. Hopefully you did what I did as I was writing the solutions and did an eye check to make sure the solution made sense.

▶ In this case, we need to be more careful about the check because the issue of domain is now a concern. When $x=8$, both $\log_6(x+4)$ and $\log_6(x-5)$ are defined, so we know that $x=8$ is part of the solution.

▶ However, neither $\log_6(x+4)$ nor $\log_6(x-5)$ is defined when $x=-7$, so this represents an extraneous root. The answer to this problem is $x=8$.

This next problem also requires the simplification of a compound statement.

▶ Solve $\log_8(x+4) - \log_8(x-12) = \dfrac{1}{3}$.

▶ Combine the left side into a single logarithm: $\log_8\left(\dfrac{x+4}{x-12}\right) = \dfrac{1}{3}$.

▶ Create the exponential equation: $\dfrac{x+4}{x-12} = 8^{\frac{1}{3}}$

$\dfrac{x+4}{x-12} = 2 \Rightarrow x+4 = 2x-24$

▶ Therefore, $x = 28$.

Not all logarithmic equations that generate quadratic equations will contain an extraneous root.

▶ Solve $2\log_4(x+2) - \log_4(x-1) = 2$.

▶ Simplify the left side of the equation into a single logarithmic statement.

$$2\log_4(x+2) - \log_4(x-1) = 2$$

$$\Rightarrow \log_4(x+2)^2 - \log_4(x-1) \Rightarrow \log_4\left(\dfrac{(x+2)^2}{x-1}\right) = 2$$

▶ Create the exponential equation: $\dfrac{(x+2)^2}{x-1} = 4^2$

$\dfrac{(x+2)^2}{x-1} = 4^2 \Rightarrow x^2 + 4x + 4 = 16x - 16$

$\Rightarrow x^2 - 12x + 20 = 0 \Rightarrow (x-10)(x-2) = 0 \Rightarrow x = 10, 2$

▶ Check both values of x to see that they both solve the original equation.

This next example appears to be more complicated, but really isn't!

EXAMPLE

▶ Solve $\log_6(x + 7) + \log_6(x + 1) - 2\log_6(x - 5) = 1$.

▶ Combine the left-hand side of the equation to a single logarithmic statement: $\log_6\left(\dfrac{(x + 7)(x + 1)}{(x - 5)^2}\right) = 1$

▶ Create the exponential equation: $\dfrac{(x + 7)(x + 1)}{(x - 5)^2} = 6^1$

$$(x + 7)(x + 1) = 6(x - 5)^2 \Rightarrow x^2 + 8x + 7 = 6x^2 - 60x + 150$$

$$\Rightarrow 5x^2 - 68x + 143 = 0$$

▶ Use the quadratic formula to solve the equation:

$$x = \frac{68 \pm \sqrt{68^2 - 4(5)(153)}}{10} = \frac{68 \pm \sqrt{1764}}{10} = \frac{68 \pm 42}{10}$$

▶ This yields x = 11, 2.6. We need to reject x = 2.6 because it is not in the domain of $\log_6(x - 5)$. Therefore, the solution to the problem is $x = 11$.

It is now time to look at the solution of some exponential equations that require the use of logarithms–and a calculator.

EXAMPLE

▶ Solve: $4^x = 39$.

▶ Thirty-nine is clearly not a power of four, nor are 4 and 39 powers of some common base. Therefore, we use logarithms to solve the problem. *I want to make this very clear.* Your calculator can work in any base you want it to. I will do this problem first with a base four logarithm and then with a natural logarithm. Why use the natural logarithm? Many of the problems you will encounter from this point forward in your mathematical career will involve Euler's number, so this is a chance for you to get used to working with natural logarithms.

$$4^x = 39 \Rightarrow \log_4(4^x) = \log_4(39)$$
$$\Rightarrow x\log_4(4) = \log_4(39)$$
$$\Rightarrow x(1) = \log_4(39)$$
$$\Rightarrow x = 2.6427$$

$$4^x = 39 \Rightarrow \ln(4^x) = \ln(39)$$
$$\Rightarrow x\ln(4) = \ln(39) \Rightarrow x = \frac{\ln(39)}{\ln(4)}$$
$$\Rightarrow x = 2.6427$$

A more complicated equation might look like this.

▶ Solve for x: $3.6e^{0.0235x} + 25 = 41.4$. Round the results to the nearest hundredth.

▶ Subtract 25 from both sides of the equation: $3.6e^{0.0235x} = 16.4$

▶ Divide by 3.6 but do not use your calculator to do the division. Leave it as it is: $e^{0.0235x} = \frac{16.4}{3.6}$

▶ Take the logarithm of both sides of the equation: $\ln(e^{0.0235x}) = \ln\left(\frac{16.4}{3.6}\right)$

▶ Apply the rule about powers: $0.0235x\ln(e) = \ln\left(\frac{16.4}{3.6}\right)$

▶ Recall that $\ln(e) = 1$. Solve for x: $x = \frac{\ln\left(\frac{16.4}{3.6}\right)}{0.0235} = 64.52$

These next problems cannot be solved algebraically. Consequently, we'll need to use our graphing utility to solve them.

EXAMPLE

▶ Solve $e^x = x + 2$.

▶ Graph each of these problems in your graphing utility.

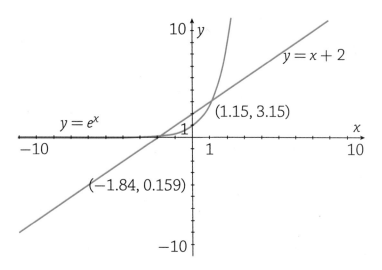

▶ Rounded to the nearest hundredth, the solution is $x = -1.84, 1.15$.

This next problem is a bit more complicated than it looks.

EXAMPLE

▶ Solve $2^x = x^2$.

▶ Before you reach for your graphing utility, take a look at the equation. Is it not obvious that $x = 2$ is a solution? The question is, is this the only solution? There are two possible ways to proceed, each of which has a potential error.

▶ Those with a certain amount of "number sense" will recognize that $x = 4$ because $2^4 = 4^2 = 16$ and will stop. This does not allow for irrational answers.

▶ There are also those who will automatically graph the functions.

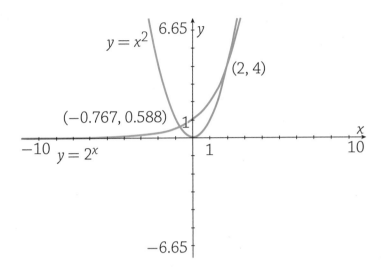

▶ These people will find the value of x that leads to the irrational solution but are guilty of accepting the default window as the appropriate window and do not see other solutions. Rather than having the window for the vertical being $[-6.65, 6.65]$, try $[-2, 50]$.

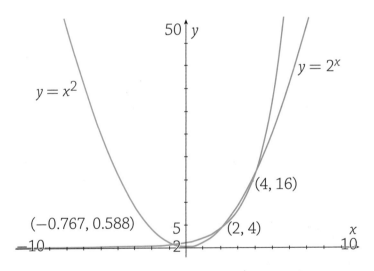

▶ They now have all three solutions. How were they to know what to do? In the default window, the exponential function lies "inside" the parabola for values of $x > 2$. However, all exponential functions with a base larger than one will eventually grow at a faster rate than any polynomial.

Let's look at another one.

▶ Solve $\log(x + 5) = x + 2$.

▶ There is not a typo in this problem, as it involves the common log. Enter each side of the equation into your graph editor and take a look at the graph.

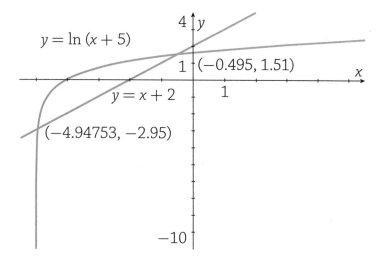

$y = \ln(x + 5)$

$(-0.495, 1.51)$

$y = x + 2$

$(-4.94753, -2.95)$

▶ We find the solution to be $x = -4.948, -0.495$ (answers rounded to the nearest thousandth).

▶ Keep in mind that as incredibly powerful as the graphing calculator is, its ability to hold onto a large number of decimal places is limited. For example, one of the solutions to the equation $\log(x + 5) = x - 2$ is $(-5, -7)$. The problem is that -5 is not in the domain of $\log(x + 5)$.

▶ So, although we know a second solution exists near $x = -5$, we still struggle to find it with the tools that we currently have at our disposal!

Applications of Exponential and Logarithmic Functions

We began the study of exponential functions with a discussion of compound interest. Let's take a look at a few of these problems. The basic formula for compound interest takes $P invested at $r\%$ compounded n times per year for a total of t years. The amount, A, in the account at the end of this period is given by the equation $A = P\left(1 + \dfrac{r}{n}\right)^{nt}$. The exponent represents the number of interest periods that occur during the given time period.

EXAMPLE

▶ If $1,000 is invested at 1.2% compounded quarterly for 5 years, how much money will be in the account at the end of that time?

▶ Plug the numbers into the formula to get $A = 1,000\left(1 + \dfrac{0.012}{4}\right)^{20} = $1,083.11.

This next example also involves compound interest but also highlights the need for logarithms.

EXAMPLE

▶ Eleanor deposits $400 into an account that pays 1.6% interest, compounded quarterly. How much time is needed before she will have $600 in his account?

▶ The equation for this problem is $600 = 400\left(1 + \dfrac{0.016}{4}\right)^{4t} = 400(1.004)^{4t}$. There is no algebraic method for solving for the exponent in a case like this without exponents.

▶ Divide by 400: $\dfrac{600}{400} = \dfrac{3}{2} = (1.004)^{4t}$

▶ Take the logarithm of both sides of the equation:

$$\ln\left(\frac{3}{2}\right) = \ln(1.004)^{4t} = 4t\ln(1.004)$$

▶ Solve for t: $t = \dfrac{\ln\left(\dfrac{3}{2}\right)}{4\ln(1.004)} = 25.3922$

▶ It would take more than 25 years for the money to grow by 50%.

This last example leads to a very interesting question.

▶ As you might well imagine, Eleanor does not want to wait 25.4 years. She would like to have the money in 3 years. What annual rate of interest, compounded quarterly, will she need to achieve this?

▶ The equation for this problem is $600 = 400\left(1 + \dfrac{r}{4}\right)^{12}$.

▶ Divide by 400: $\dfrac{3}{2} = \left(1 + \dfrac{r}{4}\right)^{12}$.

▶ Take the 12th root of both sides of the equation: $\left(\dfrac{3}{2}\right)^{\frac{1}{12}} = 1 + \dfrac{r}{4}$

▶ Subtract 1 and multiply by 4: $4\left(\left(\dfrac{3}{2}\right)^{\frac{1}{12}} - 1\right) = r$

▶ Compute: $r = 0.137464$

▶ Eleanor would need an interest rate of 13.75% to achieve her goal.

For the vast majority of us, the terms of the transactions we have with financial institutions will charge interest in time frames such as monthly, quarterly, semiannually, or annually. There are instances when the rate of interest is compounded continuously. What does that mean? If the amount of interest is computed by the expressions $\left(1 + \dfrac{r}{n}\right)^{nt}$, then we are

dealing with the situation that $n \to \infty$ so that $\left[\left(1 + \dfrac{r}{n}\right)^{n}\right]^{t} \to e^{rt}$, where

t is measured in years. The formula for the amount, A, in an account when P dollars are invested for t years at $r\%$ compounded continuously is $A = Pe^{rt}$.

EXAMPLE

▶ If $1,000 is deposited into an account that pays 1.2% compounded continuously for 5 years, how much money will be in the account after 5 years?

▶ The amount will be $1,000e^{(0.012)(5)} = 1,000e^{0.06} = \$1,061.84$.

During the 5 years, compounding continuously gained the depositor an extra $0.10.

EXAMPLE

▶ Eleanor deposits $400 into an account that pays 1.6% interest, compounded continuously. How much time is needed before she will have $600 in her account?

▶ The equation is $600 = 400e^{0.016t}$.

▶ Divide by 400: $e^{0.016t} = \dfrac{3}{2}$

▶ Take the logarithm of both sides of the equation: $\ln(e^{0.016t}) = \ln\left(\dfrac{3}{2}\right)$

▶ Simplify: $0.016t = \ln\left(\dfrac{3}{2}\right)$

▶ Solve: $t = \dfrac{\ln\left(\dfrac{3}{2}\right)}{0.016} = 25.3416$ years

Compounding continuously didn't really speed up the process very much. In the financial world, continuous compounding is used when large corporations borrow money for a short period of time.

Compound interest is an important application of exponential functions. Here are a few others.

▶ Estimates for the world population since 1995 are given by the equation $P(t) = 5344e^{0.012744t}$, where P is measured in millions of people and t is the number of years since 1990. According to this model, when will the world's population reach 8 billion?

▶ The number 8 billion is 8000 million. Substitute 8000 for $P(t)$ to get $8000 = 5344e^{0.012744t}$.

▶ Divide by 5344: $e^{0.012744t} = \dfrac{8000}{5344}$

▶ Take the logarithm of both sides: $\ln(e^{0.012744t}) = \ln\left(\dfrac{8000}{5344}\right)$

▶ Simplify: $0.012744t = \ln\left(\dfrac{8000}{5344}\right)$

▶ Solve: $t = \dfrac{\ln\left(\dfrac{8000}{5344}\right)}{0.012744} = 31.66$

▶ The model predicts the population of the world will reach 8 billion during 2026. (The World Population Review cited the U.S. Census Bureau statistics that the world population in June 2019 was 7.7 billion people. This model might need to be updated.)

Archeologists use carbon dating to estimate the age of artifacts they find.

▶ In living organic material, the ratio of the radioactive carbon-14 isotope to the content of the nonradioactive carbon-12 isotope is about $1{:}10^{12}$. When the organic material dies, the nonradioactive isotope remains constant, but the radioactive isotope begins to decay with a half-life of 5730 years. The ratio of radioactive material present after the death of the organic material to the organic material is given by $R = \dfrac{1}{10^{12}} e^{\frac{-t}{8267}}$. An archeologist finds the remains of a person from an ancient civilization and determines that the ratio in question is $1{:}10^8$. How long ago did the person die? Answer to the nearest hundred years.

▶ Substitute $\dfrac{1}{10^8}$ for R: $\dfrac{1}{10^8} = \dfrac{1}{10^{12}} e^{\frac{-t}{8267}}$

▶ Multiply by 10^{12}: $10^4 = e^{\frac{-t}{8267}}$

▶ Take the logarithm of both sides: $\ln\left(e^{\frac{-t}{8267}}\right) = \ln(10^4)$

▶ Simplify: $\dfrac{-t}{8267} = \ln(10^4)$

▶ Solve: $t = -8267\ln(10^4) = 76{,}100$

▶ The person died more than 76,000 years ago.

Logistical curves apply when there is a limit to the amount of growth or decay.

EXAMPLE

▶ A student returns to a college campus of 5000 students from vacation and has a contagious flu virus. The spread of the virus is modeled by the equation $f(t) = \dfrac{5000}{1 + 4999e^{-0.8t}}$, where $f(t)$ is the number of people with the flu and t is the number of days after the student returned to campus. The college will cancel classes when 40% of the students have the flu. When will the classes be canceled due to the flu?

▶ The number $\dfrac{f(t)}{5000}$ represents the proportion of the students who have the flu virus.

▶ Therefore, we have $0.4 = \dfrac{1}{1 + 4999e^{-0.8t}}$.

▶ Multiply: $0.4(1 + 4999e^{-0.8t}) = 1$

▶ Divide by 0.4: $1 + 4999e^{-0.8t} = 2.5$

▶ Isolate the exponential expression: $e^{-0.8t} = \dfrac{1.5}{4999}$

▶ Take the logarithm of both sides of the equation:

$\ln(e^{-0.8t}) = \ln\left(\dfrac{1.5}{4999}\right)$

▶ Simplify: $-0.8t = \ln\left(\dfrac{1.5}{4999}\right)$

▶ Solve: $t = \dfrac{\ln\left(\dfrac{1.5}{4999}\right)}{-0.8} = 10.1$

▶ Classes will be cancelled in 10 days because 40% of the student population are expected to come down with the flu.

Isaac Newton determined that the rate with which an object heats or cools is proportional to the difference between the temperature of the object and the temperature of the environment. One important application of Newton's work can be found in cooking.

One of the many challenges of cooking Thanksgiving dinner is determining when the turkey will be ready to eat.

▶ By the time a 14-pound turkey is taken out of the refrigerator, stuffed, and put into the 325° oven, the temperature of the turkey is approximately 40°. The temperature of the turkey t hours after it is put into the oven is given by $T(t) = 325 - 285e^{-0.228t}$. Dinner will be ready 30 minutes after the temperature of the turkey reaches 180°. If the turkey is put into the oven at 1:00 p.m., when will it be time to eat?

▶ Set $T(t)$ to 180: $180 = 325 - 285e^{-0.228t}$

▶ Isolate the exponential expression: $e^{-0.228t} = \dfrac{145}{285}$

▶ Take the logarithm of both sides of the equation and simplify:

$$-0.228t = \ln\left(\frac{145}{285}\right)$$

▶ Solve: $t = \dfrac{\ln\left(\dfrac{145}{285}\right)}{-0.228} = 2.96$

▶ The turkey will come out of the oven is approximately 3 hours, so dinner will be ready at 4:30 p.m. *Bon appétit!*

pH values are examples of logarithmic functions. pH measures the concentration of the hydrogen ions in a solution, [H+]. pH is computed as the negative logarithm of [H+], $pH = -\log([H+])$. A solution with a high concentration of [H+] is called an acid, while a solution with a low concentration is called a base. A solution with pH = 7 is called a neutral solution.

▶ The pH for lemon juice is 2.4. Determine the value of the hydrogen ion concentration in lemon juice.

▶ $2.4 = -\log([H^+])$

▶ Multiply by –1 and change the equation to an exponential equation:
$[H^+] = 10^{-2.4} = 0.003981$.

The planet Earth is like a living organism and it vibrates. The rates of vibration are not constant from one location to the next. The Richter scale, used to measure earthquakes, compares the ratio of the intensities of the vibrations during the earthquake to the normal vibrations of the Earth at the given location. That is, $R = \log\left(\dfrac{v_Q}{v_N}\right)$, where v_Q represents the rate of vibration during the earthquake and v_N represents the rate of vibration during normal times.

EXAMPLE

▶ On July 6, 2019, the residents of Southern California experienced an earthquake that registered 7.2 on the Richter scale. By what factor did the intensity of the vibration of the Earth change in that region during the earthquake?

▶ $7.2 = \log\left(\dfrac{v_Q}{v_N}\right) \Rightarrow 10^{7.2} = \dfrac{v_Q}{v_N} \Rightarrow v_Q = 10^{7.2}v_N = 15,848,931.9v_N$

The vibrations were almost 16 million times the normal hum of the Earth.

Sound is measured in decibels. As with the Richter scale, the decibel measurement is a comparative measurement. The formula for computing the number of decibels is $db = 10\log\left(\dfrac{I}{I_0}\right)$ where I_0 is the weakest sound that the human ear can hear.

EXAMPLE

▶ Two dishwashers are tested and the sound they make while in operation is measured. The first dishwasher has a decibel rating of 52, while the second has a decibel rating of 60. How much nosier is the second dishwasher compared to the first?

▶ We'll first determine the intensity of the sound for each dishwasher in comparison to I_0.

▶ The first dishwasher: $db = 52$

$$\Rightarrow 52 = 10\log\left(\dfrac{I}{I_0}\right) \Rightarrow 5.2 = \log\left(\dfrac{I}{I_0}\right) \Rightarrow \dfrac{I}{I_0} = 10^{5.2} \Rightarrow I = 10^{5.2}I_0$$

▶ The second dishwasher: $db = 60$

$$\Rightarrow 60 = 10\log\left(\dfrac{I}{I_0}\right) \Rightarrow 6 = \log\left(\dfrac{I}{I_0}\right) \Rightarrow \dfrac{I}{I_0} = 10^{6} \Rightarrow I = 10^{6}I_0$$

▶ The ratio of these intensities is $\dfrac{10^6}{10^{5.2}} = 10^{0.8} = 6.3$. The second dishwasher is more than 6 times louder than the first dishwasher.

EXERCISES

EXERCISE 6-1

Given $\log_b(2) = x$, $\log_b(3) = y$, and $\log_b(5) = z$, express each of the following in terms of x, y, and z.

1. $\log_b(45)$

2. $\log_b(144)$

3. $\log_b(100)$

4. $\log_{10}(3)$

5. $\log_3(10)$

Solve each of the equations in questions 6 to 25.

6. $9^{2x} = 243$

7. $4^{3x-5} = 128$

8. $8(27^{x-1}) = 72$

9. $\dfrac{2}{3}(6^{x+4}) = 144$

10. $8^{3-5x} = 16^{2x-7}$

11 $25^{x+4} = 125^{x-3}$

12. $4^{x-5} = 78$

13. $12^{2x-1} = 1000$

14. $e^{3.1x} = 178$

15. $e^{-0.28x} = 0.946$

16. $85 + 4.1e^{-0.23x} = 89.6$

17. $58 - 4.1e^{0.13x} = 49.6$

18. $(7^{3x^2})(7^{2x}) = 7^5$

19. $(4^{x^2})(32^3) = 128^{-x}$

20. $\log_2(x - 4) = 7$

21. $\log_8(2x + 1) = 4.3$

22. $\ln(4 - 5x) = -0.145$

23. $\log_2(2x + 2) + \log_2(5x + 1) = 7$

24. $\log(x + 1) + \log(9x - 2) = 2$

25. $\log_5(2x^2) - \log_5(x - 2) = 2$

EXERCISE 6-2

Solve each of the applications in the following questions.

1. How much money will be in an account after 10 years if $2500 is invested at 2% compounded quarterly?

2. How much money will be in an account after 10 years if $2500 is invested at 2% compounded continuously?

3. Aleks will need $2500 in 5 years so that she can have a down payment for a purchase. How much should Aleks invest today at 3.6% compounded monthly in order to meet her goal?

4. Aleks will need $2500 in 5 years so that she can have a down payment for a purchase. If she currently has $2000 to invest, at what rate does Aleks need to invest today if interest is compounded continuously in order to meet her goal?

5. The population of a small city was 540,000 in 2020. If the city has been experiencing an annual growth of 2.1%, in what year will the population reach 600,000?

6. The population of a town is given by the equation $P(t) = \dfrac{20}{1 + e^{-0.02t}}$, where P is measured in thousands and t measures the number of years after 2010. During which year will the population reach 15,000?

7. What is the hydrogen ion concentration of blood, which has a pH value of 7.4?

8. Doctors know that the amount of a certain drug remaining in a body decreases by 20% per hour. The equation that can be used to determine the amount, A, in milligrams, of a 10-milligram dose of the drug remaining in the body after t hours is $A(t) = 10(0.8)^t$. Find to the nearest tenth of an hour how long it takes for half the drug dose to be left in the body.

9. A 10-pound roast is at room temperature when it is put into the 325° oven. The temperature of the roast t hours after it is put into the oven is given by $R(t) = 325 - 265e^{-0.25t}$. Dinner will be ready 30 minutes after the temperature of the roast reaches 165°. If the roast is put into the oven at 4:00 p.m., when will it be time to eat?

10. The sound generated by a lawn mower has a reading of 100 decibels, while the sound generated by a garbage disposal has a reading of 80 decibels. How much louder is the lawn mower than the garbage disposal?

Flashcard App

Sequences and Series

MUST KNOW

 Geometric sequences are series of numbers in which successive terms have a common ratio. Geometric sequences have a number of practical uses, such as with mortgages and IRAs.

 Mathematical induction is a proof technique—one that's like knocking over a series of dominoes!

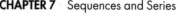

equences and series are vital elements in calculus. Newton applied these concepts in the development of the derivative, something you'll see in the last chapter. His work with series is used in electronic devices to compute the values' logarithmic and trigonometric expressions.

Sequences

A sequence is a list of numbers or terms that usually form some discernible pattern. For example:

- a, b, c, d, ...

- 1, 5, 9, 13, ...

- OTTFFSSENT . . .

- 1, –2, 4, –8, 16, . . .

The first sequence is simply the letters of the alphabet, while the second is more mathematical, as it starts with 1 and increases by four from one term to the next. The fourth sequence starts with 1 and multiplies by negative 2 to get successive terms. Do you recognize the third sequence? Here's a hint: 1, 2, 3, 4, 5, 6, 7, 8, 9, 10... The letters represent the first letter of each of the first 10 counting numbers.

A famous sequence that uses a slightly different pattern is the Fibonacci sequence:

> 1, 1, 2, 3, 5, 8, 13, 21, 34, 55, ...

This sequence starts with a pair of 1s and then every term after is found by adding the previous two terms. Fibonacci identified this sequence while raising rabbits when he was a boy. The important aspect of this sequence for us at this time is that it indicates we have a great deal of latitude in how we define the rule for creating sequences.

At this time we will concentrate only on sequences that have a numerical pattern. Sequences are defined as the range of a function that uses the set of positive integers as its input. Given this, it is natural to talk about the first term, second term, and so on until we reach the arbitrary nth term that defines the sequence. If we use the variable a to represent the terms in the sequence, the first term is a_1, the second term is a_2, and the nth term is a_n.

EXAMPLE

▶ Write the first five terms of the sequence defined by $a_n = 5n + 1$.

▶ The terms are $5(1) + 1, 5(2) + 1, 5(3) + 1, 5(4) + 1$, and $5(5) + 1$, which turn out to be 6, 11, 16, 21, and 26.

Sequences such as this example are called **arithmetic sequences** because the terms in the sequence are found by adding the same constant to go from one term to the next. The defining rule for arithmetic sequences will always be a linear function.

EXAMPLE

▶ If the fifth term in an arithmetic sequence is 45 and the twelfth term is 94, determine the value of the fiftieth term.

▶ Let the rule for this sequence be defined as $a_n = mn + b$, a basic linear function. Given $a_5 = 45 = 5m + b$ and $a_{12} = 94 = 12m + b$, solve the system of equations to determine that $m = 7$ and $b = 10$.

▶ Now that we know $a_n = 7n + 10$, we can see that $a_{50} = 7(50) + 10 = 360$.

Sequences such as 1, –2, 4, –8, 16, ... are called **geometric sequences**. In a geometric sequence, successive terms have a common ratio and the function that generates the geometric sequence is exponential. There is an important difference between the exponential functions studied in Chap. 6 and how we are using them here. In general, the domain of an exponential function is the set of real numbers, while in this context the domain is the set of positive integers. Given this, we can allow a negative number to be the base of the generating function. We will never encounter the expression $(-2)^{\frac{1}{2}}$ in a geometric sequence but could in the general application of exponential functions.

▶ Write the first five terms of the geometric sequence defined by the rule $a_n = 4(3)^{n-2}$.

▶ The first five terms of the geometric sequence are $4(3)^{1-2}$, $4(3)^{2-2}$, $4(3)^{3-2}$, $4(3)^{4-2}$, and $4(3)^{5-2}$.

▶ When the computations are completed, the terms are $\dfrac{4}{3}$, 4, 12, 36, and 108.

Compound interest is an example of a geometric sequence. If $1,000 is invested in an account with a periodic rate of interest of 0.5%, the amount of money in the account after the first interest period is $1,000 + 1,000(0.005) = 1,000(1 + 0.005) = 1,000(1.005)$. At the end of the second interest period, the amount of money in the account will be $1,000(1.005) + 1,000(1.005)(0.005) = 1,000(1.005)(1 + 0.005) = 1,000(1.005)^2$. You can see that this sequence has the common factor 1.005 so that the general term for the sequence is $a_n = 1,000(1.005)^n$.

▶ If the sixth term of a geometric sequence is 20,480 and the tenth term is 5,242,800, determine the value of the third term.

▶ Let the rule for this geometric sequence be defined as $a_n = a_1(r)^{n-1}$. The sixth term is $20,480 = a_1(r)^5$ and the tenth term is

$5{,}242{,}800 = a_1(r)^9$. We can solve this system of equations by dividing the terms in the second equation by the terms in the first:

$$\frac{5{,}242{,}800}{20{,}480} = \frac{a_1(r)^9}{a_1(r)^5} \Rightarrow 256 = r^4.$$ Solve this equation to determine

that $r = \pm 4$. Substitute $r = 4$ into either equation to find $a_1 = 5$. Substitute $r = -4$ into either equation to find $a_1 = -5$.

▶ In either case, the third term of this sequence is $a_3 = 5(4)^3 = 320$.

Not all numerical sequences need be arithmetic or geometric. Take a moment to write the first five terms of the sequences $a_n = 5 + 2^n$, $a_n = 5n + 2^n$, and $a_n = \dfrac{n+1}{n+2}$.

$$a_n = 5 + 2^n : 7, 9, 13, 21, 37$$

$$a_n = 5n + 2^n : 7, 14, 23, 36, 57$$

$$a_n = \frac{n+1}{n+2} : \frac{2}{3}, \frac{3}{4}, \frac{4}{5}, \frac{5}{6}, \frac{6}{7}$$

Finally, there are sequences that are defined based on a previous value of the sequence. Such definitions are called **recursive** (although in the world of computing, they are called *iterative*). It is usually the case that the first term in a recursive sequence is defined by a constant.

EXAMPLE

▶ Write the first five terms of the sequence defined by $a_1 = 1{,}000$ and $a_n = 1.005 a_{n-1} (n \geq 2)$.

▶ Observe that the definition requires that we define at the first term, if not more terms, when to use the recursive piece of the formula. The first five terms are $1{,}000$, $1{,}000(1.005)$, $1{,}000(1.005)^2$, $1{,}000(1.005)^3$, and $1{,}000(1.005)^4$. These are just the terms of the example we used for compound interest.

▶ One could argue that this is a better equation for compound interest than the exponential function because it distinctly says that whatever value available at the beginning of the term, the value at the end will be 1.005 times as much.

▶ I did not say that this formula would make it easier to compute the value of the account after 20 periods. In this style, we would need to compute the amount in the account after the first 19 periods as well.

Observe that the recursive formula for this last example tells us to multiply the last value by 1.005. Another way of saying this is that the ratio of consecutive terms will be a constant. That automatically tells us the sequence is geometric. In the same way, a sequence such as $a_1=12$; $a_n=a_{n-1}+4$ $(n \geq 2)$ must be arithmetic because there is a constant difference between successive terms. However, the sequence $a_1=12$; $a_n=a_{n-1}+n$ $(n \geq 2)$ is not arithmetic because the difference between terms will always increase by 1 due to the expression n being the value added.

EXAMPLE

▶ Write the first five terms in the sequence defined by $a_1 = 1{,}000$; $a_n = 1.005a_{n-1} + 1{,}000$ $(n \geq 2)$.

▶ This looks like the compound interest problem again (and, in fact, is an important variation of it). We'll take a closer look at this problem later in the chapter. The first five terms are

1,000, 1.005(1,000) + 1,000, 1.005(1.005(1,000) + 1,000) + 1,000, 1.005(1.005(1.005(1,000) + 1,000) + 1,000, 1.005(1.005(1.005(1.005(1,000) + 1,000) + 1,000) + 1,000) + 1,000.

▶ Let's rewrite these terms in a more compact notation.

1,000,

$1.005(1{,}000) + 1{,}000,$

$1.005^2(1{,}000) + 1.005(1{,}000) + 1{,}000,$

$1.005^3(1{,}000) + 1.005^2(1{,}000) + 1.005(1{,}000) + 1{,}000,$

$1.005^4(1{,}000) + 1.005^3(1{,}000) + 1.005^2(1{,}000) + 1.005(1{,}000) + 1{,}000$

As mentioned, there will be more about this in a little while.

The Fibonacci sequence can be defined recursively as $a_1 = 1; a_2 = 1;$ $a_n = a_{n-1} + a_{n-2}$ $(n \geq 3)$. The first two terms are defined by constants, but the third term is $a_3 = a_2 + a_1 = 1 + 1 = 2$ and the fourth term is $a_4 = a_3 + a_2 = 2 + 1 = 3$.

EXAMPLE

▶ Write the first five terms of the sequence defined by
$a_1 = 2; a_2 = 3; a_n = 5a_{n-1} - 3a_{n-2}$ $(n \geq 3)$.

▶ The first two terms are easy: 2 and 3.

▶ The third term is $a_3 = 5a_2 - 3a_1 = 5(3) - 3(2) = 9$.

▶ The fourth term is $a_4 = 5a_3 - 3a_2 = 5(9) - 3(3) = 36$.

▶ The fifth term is $a_5 = 5a_4 - 3a_3 = 5(36) - 3(9) = 153$.

Summation Notation

Consider this set of data:

i	1	2	3	4	5	6	7	8	9	10
x_i	78	81	29	38	103	91	12	54	62	81

The mean of the data is found by adding all the data points and dividing by the number of data values. That is, $\bar{x} = \dfrac{\displaystyle\sum_{i=1}^{10} x_i}{10}$.

The uppercase Greek letter sigma (Σ) has long been used in mathematics to indicate a summation process.

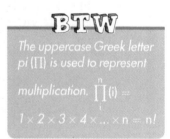

BTW

The uppercase Greek letter pi (Π) is used to represent multiplication. $\displaystyle\prod_{i}^{n}(i) = 1 \times 2 \times 3 \times 4 \times ... \times n = n!$

We will apply the summation notation as we work with adding the terms generated by a sequence.

EXAMPLE

▶ Compute the sum of the first five terms of the sequence defined by $a_n = 5n + 1$.

▶ $\displaystyle\sum_{n=1}^{5}(5n + 1) = 6 + 11 + 16 + 21 + 26 = 80$

This next example highlights another skill that is worthwhile in investigating sequences and series.

▶ Write $15 + 19 + 23 + 27 + 31 + 35 + 39 + 43 + 47 + 51$ using summation notation.

▶ There is a technique known as successive differences that gives a clue as to the nature of the function defining the terms in a sequence. (This technique is actually an application of the derivative in calculus but works well here.) If the first successive differences are constant, then the defining function is linear. If the second successive differences are constant, the defining function is quadratic. This pattern continues for higher-order polynomials.

$$
\begin{array}{cccccccccccc}
15 & & 19 & & 23 & & 27 & & 31 & & 35 & & 39 & & 43 & & 45 & & 47 & & 51 \\
& \vee & & \vee & & \vee & & \vee & & \vee & & \vee & & \vee & & \vee & & \vee & & \vee \\
& 4 & & 4 & & 4 & & 4 & & 4 & & 4 & & 4 & & 4 & & 4 & & 4
\end{array}
$$

▶ The first differences are constant, so the defining function is linear and, for the added bonus, the slope is 4. With the first term being 15, you can solve the equation $15 = 4(1) + b$ to determine the defining function is $a_n = 4n + 11$.

▶ Therefore, we can write $15 + 19 + 23 + 27 + 31 + 35 + 39 + 43 +$

$$47 + 51 = \sum_{n=1}^{10} 4n + 11.$$

Let's take a look at a slightly different one of these.

▶ Write $3 + 9 + 17 + 33 + 65 + 129 + 257$ using summation notation.

▶ Look at successive differences.

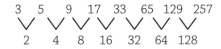

$$
\begin{array}{ccccccccccccc}
3 & & 5 & & 9 & & 17 & & 33 & & 65 & & 129 & & 257 \\
& \vee & & \vee & & \vee & & \vee & & \vee & & \vee & & \vee \\
& 2 & & 4 & & 8 & & 16 & & 32 & & 64 & & 128
\end{array}
$$

▶ The differences are not constant, but there is a pattern to them—they are all powers of 2. This is an indication that the defining function is exponential and the base of the exponential function is 2. Look at the numbers in the original expression. They are all one more than a power of 2. Do you see that the defining function must be $a_n = 2^n + 1$?

▶ We now know that $3 + 9 + 17 + 33 + 65 + 129 + 257 = \sum_{n=1}^{8} 2^n + 1$.

Arithmetic Series

A **series** is the sum of the terms in a sequence. An **arithmetic series** is the sum of the terms in an arithmetic sequence. Carl Gauss showed that the sum of the series is equal to $S_n = \dfrac{n}{2}(a_1 + a_n)$. When asked to find the sum of the first 100 counting numbers, Gauss wrote

$$S = 1 + 2 + 3 + 4 + \dots + 98 + 99 + 100$$

His stroke of genius was that he then wrote

$$S = 100 + 99 + 98 + 97 + \dots + 3 + 2 + 1$$

When he added the two equations together, the result was

$$2S = 101 + 101 + 101 + 101 + \dots + 101 + 101 + 101$$

$$2S = 100(101)$$

$$S = \frac{100}{2}(101)$$

This reasoning works for all arithmetic series.

IRL Late in the 18th century, there was a five-year old student in a K to 8 classroom in Germany. According to this story, he and his classmates were directed to find the sum of the first 100 counting numbers. Before the teacher had a chance to sit down, the five-year-old brought his slate to the teacher with the correct answer on it. His analysis of the problem was the beginning of the study of arithmetic sequences and series. The student, Carl F. Gauss, would grow up to be one of the world's most famous mathematicians.

Using the formula $a_n = dn + b$, where d is the common difference of the terms in the arithmetic sequence and b is a constant, the first term of the sequence is $a_1 = d + b$. We can use these two equations to write an alternative form for the sum of the arithmetic series that is reliant only on the first term. Rewrite the expression for the nth term as $a_n = d(n-1) + d + b = d(n-1) + a_1$. The sum of the first n terms of the arithmetic series can now be written as $S_n = \dfrac{n}{2}(a_1 + a_1 + d(n-1)) = \dfrac{n}{2}(2a_1 + d(n-1))$.

BTW

There may be some problems in which this method is the quicker choice for finding the sum, but it requires one more thing to memorize and I'd rather not do that!

EXAMPLE

▶ Find the sum of the first 250 terms of the sequence defined by $a_n = 12n - 7$.

▶ The first term of the series is $a_1 = 5$ and the last term is $a_{250} = 2993$.

▶ The sum of the first 250 terms is $S_{250} = \dfrac{250}{2}(5 + 2993) = 374{,}750$.

Here is an interesting application of arithmetic series to compute a sum.

EXAMPLE

Determine the value of $\displaystyle\sum_{n=101}^{1000} 3n+2$.

Yes, it is true we can first compute the values of a_{101} and a_{1000} and then determine the sum. However, we can also write

$$\sum_{n=101}^{1000}(3n+2) = \sum_{n=1}^{1000}(3n+2) - \sum_{n=1}^{100}(3n+2)$$

$$\sum_{n=101}^{1000}(3n+2) = \frac{1000}{2}(5+3002) - \frac{100}{2}(5+302)$$

$$\sum_{n=101}^{1000}(3n+2) = 1{,}488{,}150$$

This next example is easily solved by the alternative formula for the sum of the arithmetic series. The extra work needed to solve the problem with the original formula is not that difficult.

EXAMPLE

Find the sum of the first 200 terms of the series $21 + 27 + 33 + 39 + \ldots$.

Alternative form: The first term in the series is 21 and the common difference is 6. Therefore, $S_{200} = \frac{200}{2}(2(21) + 199(6)) = 123{,}600$.

Original form: The first term is 21 and the common difference is 6. Therefore, $a_1 = 21 = 6(1) + b \Rightarrow b = 15$. So $a_{200} = 200(6) + 15 = 1215$. The sum of the first 200 terms is $S_{200} = \frac{200}{2}(21 + 1215) = 123{,}600$.

We certainly have the choice of how to solve these problems!

Geometric Series

As you might easily surmise, a **geometric series** is the sum of the terms of a geometric sequence. The derivation of the formula for a geometric series is not quite as obvious as that for an arithmetic series, but it is not all that difficult.

$$S_n = a_1 + a_1 r + a_1 r^2 + a_1 r^3 + \ldots + a_1 r^{n-1}$$

Multiply both sides of the equation by r.

$$r S_n = a_1 r + a_1 r^2 + a_1 r^3 + a_1 r^4 + \ldots + a_1 r^{n-1} + a_1 r^n$$

Subtract these two equations from each other. All the terms on the right side of the equation will drop out with the exception of a_1 and $a_1 r^n$.

$$S_n - r S_n = a_1 - a_1 r^n$$

Factor:
$$S_n(1 - r) = a_1(1 - r^n)$$

Solve:
$$S_n = \frac{a_1(1 - r^n)}{1 - r}$$

Let's put this to use in an example.

▶ Find the sum of the first 30 terms of the geometric series $3 + 6 + 12 + 24 + 48 + \ldots$.

▶ The first term is 3 and the common ratio is 2. Therefore,

$$S_{30} = \frac{3(1 - 2^{30})}{1 - 2} = 3,221,225,469.$$

There is a legend that the person who developed the game of chess was asked to show the game to the king. The king was so impressed that he offered the man anything he wanted. The man asked for 1 grain of wheat for the first square on the board, double that for the second square, and continue to double the number of grains of wheat on the next square for the rest of the board. The king agreed ... until he learned how many grains of wheat he was expected to pay. The total number of grains of wheat:

$$1 + 2 + 4 + 8 + 16 + \ldots + 2^{63} = \frac{1 - 2^{64}}{1 - 2} = 18{,}446{,}744{,}073{,}709{,}551{,}615$$

Needless to say, the man who created the game was not paid.

This next example also involves a fairly large sum!

EXAMPLE

▶ Determine the sum of the first 20 terms of the series
1,200 + 600 + 300 + 150 + ...

▶ The first term is 1,200 and the common ratio is $\frac{1}{2}$, so the sum of the first 20 terms is

$$S_{20} = \frac{1{,}200\left(1 - \left(\frac{1}{2}\right)^{20}\right)}{1 - \frac{1}{2}} = \frac{78{,}643{,}125}{32{,}768}.$$

This last example raises an interesting question. The terms in the sequence get infinitesimally small as $n \to \infty$. Is there a limit to the value of the sum of those terms? That is, as $n \to \infty$, does $S_n = \dfrac{1{,}200\left(1 - \left(\frac{1}{2}\right)^n\right)}{1 - \frac{1}{2}}$

approach some finite value? The answer is yes. As $n \to \infty$, $\left(\dfrac{1}{2}\right)^n \to 0$

and $S_n = \dfrac{1{,}200\left[1 - \left(\dfrac{1}{2}\right)^n\right]}{1 - \dfrac{1}{2}} \to \dfrac{1{,}200}{1 - \dfrac{1}{2}} \to 2{,}400.$

We get the following equation for the sum of **infinite geometric series**: If $|r| < 1$ in a geometric series, the sum of the infinite number of terms, S_∞,

is $S_\infty = \dfrac{a_1}{1 - r}.$

▶ Determine the sum of $1 + \dfrac{1}{2} + \dfrac{1}{4} + \dfrac{1}{8} + \ldots.$

▶ The first term is 1 and the common ratio is $\dfrac{1}{2}$, so $S_\infty = \dfrac{1}{1 - \dfrac{1}{2}} = 2.$

The statement about the sum of an infinite geometric series states that the absolute value of the common ratio has to be less than 1. We can have common ratios that are negative.

▶ Determine the sum of $1 - \dfrac{1}{2} + \dfrac{1}{4} - \dfrac{1}{8} + \ldots.$

▶ The first term is 1 and the common ratio is $\dfrac{-1}{2}$, so $S_\infty = \dfrac{1}{1 - \dfrac{-1}{2}} = \dfrac{2}{3}.$

Before we take on the financial application of the geometric series, let's take a look at two examples of compound interest that will help us understand why people in the financial world use the phrase "time is money." The first example is about a deposit of $5,000 in an interest-bearing account, and the second is a question of reaching the total of $5,000 in an interest-bearing account.

EXAMPLE

▶ If $5,000 is deposited into an account that pays 2% interest compounded quarterly, how much money will be in the account after 10 years?

▶ A timeline for this problem shows

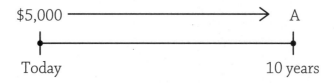

$5,000 ─────────────────────→ A

Today   10 years

▶ The value of the money in today's market is $5,000. In 10 years, that

money will be worth $5,000\left(1 + \dfrac{.02}{4}\right)^{40} \doteq 5,000(1.005)^{40} = \$6,103.97.$

This second example determines what $5,000 10 years from now is worth today under the same scenario.

▸ How much money must be deposited in an account that pays 2% interest compounded quarterly so that there will be $5,000 in the account at that time?

▸ The timeline for this problem is different.

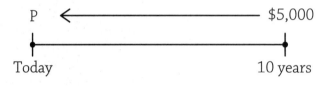

$$5,000 = P(1.005)^{40} \Rightarrow P = 5,000(1.005)^{-40} = \$4,095.69$$

There are two financial scenarios that you will have to deal with in your life (if you are not already dealing with them) and they are saving for your future (particularly retirement—don't snicker!) and paying back a loan.

Let's begin with the example of a scheduled savings plan.

▸ Mavis, age 29, decides that she is going to establish a savings plan for her future. Beginning on her 30th birthday, Mavis will deposit $1,500 into an account that pays 2% compounded quarterly. She will make additional $1,500 payments every 3 months after her 30th birthday with the last payment to be made on her 65th birthday. How much money will be in the account after the last payment is made?

▸ The first payment will collect interest for 35 years, which represents 140 interest periods. Her second payment, made 3 months after her birthday, will collect interest for 139 interest periods. The third payment will collect interest for 138 interest periods. The pattern continues for each successive payment until the last payment on her

65th birthday. This payment will not collect any interest. The amount of money in the account will be

$$A = 1,500(1.005)^{140} + 1,500(1.005)^{139} + 1,500(1.005)^{138}$$
$$+ \ldots + 1,500(1.005) + 1,500$$

▶ If you read the numbers from last payment to first payment, you can see the geometric series with $a_1 = 1,500$ and $r = 1.005$. Mavis has made 141 payments into the account. So, at age 65 she will have

$$S_{141} = \frac{1,500\left(1 - 1.005^{141}\right)}{1 - 1.005} = \$306,088.39 \text{ in her account.}$$

There are a couple of things worth noting here. First, $306,000 is a good start to a retirement fund! Second, it is usually the case that people earn more money as they have more experience, so it is not unreasonable to believe that Mavis will be able to increase the amount of her quarterly payments. The third item is mathematical in nature. We saw this problem earlier when we examined recursively defined sequences. The scenario described in this example can be modeled by $a_1 = 1,000$; $a_n = 1.005a_{n-1} + 1,000$ ($n \geq 2$).

The general equation for a savings plan with equal deposits of $P into an account that pays $r\%$ interest each period for n interest periods is

$$S_n = \frac{P\left(1 - (1 + r)^n\right)}{-r}.$$

This next problem involves paying back a loan with equal periodic payments. The two most likely cases you will someday encounter are car loans and mortgage payments.

EXAMPLE

▶ Carlita takes out a $25,000 car loan. The loan is for 5 years and the interest rate is 3.6% compounded monthly. What is the amount of her monthly payment?

▶ Carlita has 60 payments she needs to make to repay the loan. The way the process works, she will make her first payment of P, the payment will be deducted from the balance of her account, and then interest will be computed and added to the remaining balance. Her 60th payment will exactly match the balance due on her account. The timeline for this loan looks like this:

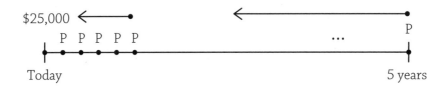

▶ We have $25,000 = P(1.003)^{-1} + P(1.003)^{-2} + P(1.003)^{-3} + \ldots + P(1.003)^{-60}$. Let's make this look more like a geometric series. We will multiply both sides of the equation by $(1.003)^{60}$ and get

$$25,000(1.003)^{60} = P(1.003)^{59} + P(1.003)^{58} + P(1.003)^{57} + \ldots + P(1.003) + P$$

▶ The right-hand side of the equation equals $\dfrac{P\left(1 - (1.003)^{60}\right)}{1 - 1.003}$. This gives the equation

$$25,000(1.003)^{60} = \dfrac{P\left(1 - (1.003)^{60}\right)}{1 - 1.003}$$

▶ Solve for P:

$$P = \dfrac{25,000(1.003)^{60}(1 - 1.003)}{\left(1 - (1.003)^{60}\right)} = \$455.91$$

In general, the amount of each payment is $P = \dfrac{A(1+r)^n(-r)}{1-(1+r)^n}$, where A is the amount borrowed, r is the periodic rate of interest, and n is the number of interest payments.

The difference between a car loan and a mortgage is the amount of money borrowed, the number of years to repay, and, possibly, the rate of interest.

EXAMPLE

▶ The Ming-Suns take out a mortgage for their new home. If they borrow $275,000 for 30 years at a rate of 3.75%, what is the amount of each payment?

▶ Using $A = 275,000$, $n = 360$, and $r = \dfrac{0.0375}{12} = 0.003125$, each payment will be

$$P = \frac{275,000(1.003125)^{60}(1-0.003125)}{1-(1.003125)^{60}} = \$1{,}273.57$$

Mathematical Induction

Mathematical induction is often likened to knocking over dominoes that have been stacked one behind the other. The thinking goes like this: I can knock over the first domino; I assume that when a domino falls, the next domino in line will also fall. Therefore, all the dominoes will fall.

With mathematical induction, we apply three steps.

- **Step 1.** Prove the statement is true for the first case.

- **Step 2.** Assume the statement is true for an arbitrary case $n = k$.

- **Step 3.** Prove the statement is true for the next case, $n = k + 1$.

Keep in mind that when executing Steps 1 and 3, you cannot move terms from one side of the equation to the other. The process of moving terms across the equal sign assumes that the items are equal. Consequently, the proofs are often written in column form with a vertical bar, rather than the equal sign, separating the columns.

EXAMPLE

▶ Prove $3^n - 1$ is an even number.

▶ Step 1. Prove for $n = 1$: $3^1 - 1$ is 2, and that is an even number.

▶ Step 2. Assume for $n = k$: $3^k - 1 = 2m$, where m is an integer. (We learned a long time ago that any integer multiplied by 2 must be even.)

▶ Step 3. Prove for $n = k + 1$: Show that $3^{k+1} - 1$ is equal to $2p$, where p is some integer.

▶ Factor 3 from 3^{k+1}: $3^{k+1} - 1 \Rightarrow 3(3^k) - 1$.

▶ In Step 2, we assumed that $3^k - 1 = 2m$, so $3^k = 2m + 1$.

▶ Therefore, $3(3^k) - 1 \Rightarrow 3(2m + 1) - 1 = 6m + 3 - 1 = 6m + 2 = 2(3m + 1)$. Since 2 is a factor of this number, the number must be even. The proof is complete.

Let's look at a few other interesting examples.

EXAMPLE

▶ Prove that the sum of the first n odd integers is n^2.

▶ That is to say, $1 + 3 + 5 + 7 + \ldots + 2n - 1 = n^2$.

▶ Step 1. Prove for $n = 1$: $1 = 1^2$.

▶ Step 2. Assume for $n = k$: $1 + 3 + 5 + 7 + \ldots + 2k - 1 = k^2$.

Step 3. Prove for $n = k + 1$: $1 + 3 + 5 + 7 + \ldots + 2k - 1 + 2k + 1 = (k+1)^2$.

Group the sum of the first k terms: $(1 + 3 + 5 + 7 + \ldots + 2k - 1) + 2k + 1$

Replace the group with k^2
(from assumption): $\qquad\qquad \Rightarrow k^2 + 2k + 1$

Factor: $\qquad\qquad\qquad\qquad \Rightarrow (k+1)^2$

The proof is complete.

The next two examples are, surprisingly, related to each other.

EXAMPLE

The triangular numbers are the sum of the first n positive integers.

Prove that the sum of the first n positive integers is $\dfrac{n(n+1)}{2}$.

Before we do the proof by induction, this picture gives a compelling geometric argument.

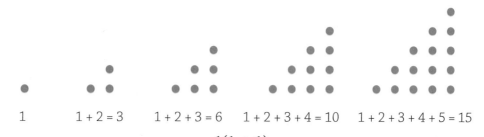

| 1 | $1 + 2 = 3$ | $1 + 2 + 3 = 6$ | $1 + 2 + 3 + 4 = 10$ | $1 + 2 + 3 + 4 + 5 = 15$ |

Step 1. Prove for $n = 1$: $\quad 1 = \dfrac{1(1+1)}{2} = 1$

Step 2. Assume for $n = k$: $1 + 2 + 3 + 4 + 5 + \ldots + k = \dfrac{k(k+1)}{2}$

▶ Step 3. Prove for $n = k + 1$: $1 + 2 + 3 + 4 + 5 + \ldots + k + (k + 1) =$

$$\frac{(k+1)((k+1)+1)}{2} = \frac{(k+1)(k+2)}{2}$$

▶ Group the first k terms: $(1 + 2 + 3 + 4 + 5 + \ldots + k) + k + 1$

▶ Replace the grouping with the result of assumption: $\Rightarrow \dfrac{k(k+1)}{2} + k + 1$

▶ Get a common denominator: $\Rightarrow \dfrac{k(k+1)}{2} + \dfrac{2(k+1)}{2}$

▶ Combine fractions and factor: $\Rightarrow \dfrac{k(k+1)+2(k+1)}{2} \Rightarrow \dfrac{(k+1)(k+2)}{2}$

▶ The proof is complete.

Let's look at one final example.

EXAMPLE

▶ Prove the sum of the cubes of the first n positive integers is $\dfrac{n^2(n+1)^2}{4}$.

▶ Did you notice that $\dfrac{n^2(n+1)^2}{4}$ is the square of the sum of the first n positive integers?

▶ Step 1. Prove for $n = 1$: $\quad 1^3 = \dfrac{(1)^2(1+1)^2}{4} \Rightarrow 1 = 1$

▶ Step 2. Assume for $n = k$: $1^3 + 2^3 + 3^3 + 4^3 + \ldots + k^3 = \dfrac{k^2(k+1)^2}{4}$

▶ Step 3. Prove for $n = k + 1$: $1^3 + 2^3 + 3^3 + 4^3 + \ldots + k^3 + (k+1)^3$

$$= \dfrac{(k+1)^2(k+2)^2}{4}$$

▶ Group the first k terms: $(1^3 + 2^3 + 3^3 + 4^3 + \ldots + k^3) + (k+1)^3$

▶ Substitute $\dfrac{k^2(k+1)^2}{4}$ from our assumption: $\dfrac{k^2(k+1)^2}{4} + (k+1)^3$

▶ Get the common denominator: $\dfrac{k^2(k+1)^2}{4} + \dfrac{4(k+1)^3}{4}$

▶ Combine fractions and factor $(k+1)^2$: $\dfrac{(k+1)^2\left(k^2 + 4(k+1)\right)}{4}$

$$= \dfrac{(k+1)^2(k^2 + 4k + 4)}{4}$$

▶ Factor trinomial: $\dfrac{(k+1)^2(k+2)^2}{4}$

▶ The proof is complete!

EXERCISES

EXERCISE 7-1

Find the next five terms in each of the sequences in questions 1 to 3.

1. 29, 52, 75, 98, ...

2. J, F, M, A, M, ...

3, 240, 360, 540, 810, ...

Find the first five terms of the sequence defined by each formula in questions 4 to 8.

4. $a_n = 40n + 30$

5. $a_n = 1800\left(\dfrac{-2}{3}\right)^{n-1}$

6. $a_1 = 8; a_n = 3a_{n-1} + 5; (n \geq 2)$

7. $a_1 = 2; a_2 = 4; a_n = a_{n-2}(a_{n-1})^2; (n \geq 3)$

8. $a_1 = 2; a_2 = 4; a_n = 4a_{n-1} + 3a_{n-2}; (n \geq 3)$

Determine the value requested in questions 9 to 15.

9. Find the 25th term of an arithmetic sequence if the 5th term is 52 and the 9th term is 84.

10. Find the 15th term of a geometric sequence in which the 3rd term is 720 and the 15th term is 6,480.

11. Evaluate $\displaystyle\sum_{n=1}^{85}(7n - 5)$.

12. Evaluate $\displaystyle\sum_{n=1}^{15}5(1.7)^{n-1}$

13. Evaluate $\displaystyle\sum_{n=1}^{50}(77-6n)$.

14. Evaluate $\displaystyle\sum_{n=1}^{15}5400\left(\frac{2}{3}\right)^{n-1}$.

15. Evaluate $\displaystyle\sum_{n=1}^{\infty}49\left(\frac{3}{10}\right)^{n-1}$.

EXERCISE 7-2

Answer these mathematical induction–related questions.
 Determine the amount of money described in questions 1 to 3.

1. Marvin, age 35, decides that he is going to establish a savings plan for his future. Beginning on his 36th birthday, Marvin will deposit $500 into an account that pays 2.4% compounded monthly. He will make additional $500 payments every month after his 36th birthday with the last payment to be made on his 65th birthday. How much money will be in the account after the last payment is made?

2. Max takes out a $35,000 car loan. The loan is for 6 years and the interest rate is 2.4% compounded monthly. What is the amount of her monthly payment?

3. The Merkels take out a mortgage for their new home. If they borrow $475,000 for 25 years at a rate of 4.5%, what is the amount of each payment?

4. Use mathematical induction to prove the sum of the squares of the first n positive integers is $\dfrac{n(n+1)(2n+1)}{6}$.

Use the results of the proofs completed by mathematical induction to answer questions 5 to 8.

5. $\displaystyle\sum_{n=1}^{250}(2n-1)$

6. $\displaystyle\sum_{n=1}^{150}(2n)$

7. $\displaystyle\sum_{n=1}^{20}(n^3)$

8. $\displaystyle\sum_{n=1}^{20}(n^3+n^2-n)$

Determine the value of each series in questions 9 and 10.

9. $125+25+5+1+\dfrac{1}{5}+\ldots$

10. $125-25+5-1+\dfrac{1}{5}-\ldots$

 **Systems of Equations and Matrices**

You have been solving systems of linear equations since Algebra I. Whether you used substitution or a linear transformation method (multiplying equations by constants and then adding equations), the goal was to reduce the problem to a single equation in a single variable. While we could continue that process involving three equations in three variables, four equations in four variables, or *n* equations in *n* variables, applying a matrix solution is a much faster method.

An Introduction to Matrices

It is not uncommon for a car dealership to have multiple stores across a geographic region. Consider the case of the C³ Brothers automobile dealership with stores in Charlotte, North Carolina; Chicago; Cincinnati; and Concord, New Hampshire. The August inventory is taken of five of its bestselling SUVs: Toyota Avalon, Lexus RX, Buick LaCrosse, Lincoln MKZ, and BMW 3 Series. The table below shows the inventory for each model at each location in August.

	Toyota Avalon	Lexus RX	Buick LaCrosse	Lincoln MKZ	BMW 3 Series
Charlotte, NC	18	17	25	10	15
Chicago	12	21	28	20	15
Cincinnati	10	17	16	13	25
Concord, NH	12	18	12	9	21

The inventory numbers for September of these same models are as shown.

	Toyota Avalon	Lexus RX	Buick LaCrosse	Lincoln MKZ	BMW 3 Series
Charlotte, NC	13	14	21	9	17
Chicago	15	20	25	18	18
Cincinnati	9	19	14	14	23
Concord, NH	11	21	7	11	21

What is the difference in the number of models available at each of the locations? Common sense tells us to compare each of the cells. Working from the notion that we take the September numbers and subtract the August numbers, we get the following.

	Toyota Avalon	Lexus RX	Buick LaCrosse	Lincoln MKZ	BMW 3 Series
Charlotte, NC	–5	–3	–4	–1	2
Chicago	3	–1	–3	–2	3
Cincinnati	–1	2	–2	1	–2
Concord, NH	–1	3	–5	2	0

We can add and subtract matrices that have the same dimensions by adding and subtracting corresponding cells. Furthermore, two matrices are said to be equal provided the matrices have the same dimensions and all corresponding cells are equal.

EXAMPLE

▶ If $\begin{bmatrix} 2x+1 & 5y-3 \\ 2x+3y & 3x-2y \end{bmatrix} = \begin{bmatrix} w & z \\ 29 & 11 \end{bmatrix}$, find the values of w, x, y, and z.

▶ Solve the system of equations

$2x+3y = 29$
$3x-2y = 11$

to determine that $x = 7$ and $y = 5$. Then $w = 2(7)+1 = 15$ and $z = 5(5)-3 = 22$.

The wholesale price of each model, the price the C³ Brothers paid for each model, is shown.

Model	Wholesale Price
Toyota Avalon	$35,800
Lexus RX	$39,750
Buick LaCrosse	$29,570
Lincoln MKZ	$36,750
BMW 3 Series	$40,750

EXAMPLE

▶ Compute the value of the SUV inventory for the month of August at each of the C^3 Brothers' four stores.

▶ For each store, multiply the number of each model by the wholesale cost of the model.

▶ Charlotte: 18(35,800) + 17(39,750) + 25(29,570) + 10(36,750) + 15(40,750) = $3,038,150

▶ Chicago: 12(35,800) + 21(39,750) + 28(29,570) + 20(36,750) + 15(40,750) = 3,438,560

▶ Cincinnati: 10(35,800) + 17(39,750) + 16(29,570) + 13(36,750) + 25(40,750) = 3,003,370

▶ Concord: 12(35,800) + 18(39,750) + 12(29,570) + 9(36,750) + 21(40,750) = 2,686,440

We will now use this example to illustrate the mathematical construct called a **matrix**. A matrix is a rectangular array of numbers. Returning to the first table on page 176 (and ignoring the column and row names), we see that the table has four rows and five columns. (Rows are read horizontally and columns vertically.) Now look at the table at the top of this page, which has five rows and one column. The number of rows and the number of columns identifies the **dimensions of a matrix**. Let **A** represent the inventory, from the *first* table on page 176, and **B** the wholesale value, from the table on page 178. The dimensions of **A** are 4 × 5 and the dimensions of **B** are 5 × 1.

$$A = \begin{bmatrix} 18 & 17 & 25 & 10 & 15 \\ 12 & 21 & 28 & 20 & 15 \\ 10 & 17 & 16 & 13 & 25 \\ 12 & 18 & 12 & 9 & 21 \end{bmatrix} \text{ and } B = \begin{bmatrix} 35{,}800 \\ 39{,}750 \\ 29{,}570 \\ 36{,}750 \\ 40{,}750 \end{bmatrix}$$

Multiply matrices **A** and **B** to get

$$AB = \begin{bmatrix} 3{,}038{,}150 \\ 3{,}438{,}560 \\ 3{,}003{,}370 \\ 2{,}686{,}440 \end{bmatrix}$$

These are the same values calculated in the solution to the first example, the total value of the inventory for each store. This tells us the process in which multiplication of matrices occurs. Starting with row 1 of the left-hand factor, multiply each element in row 1 with its corresponding element from column 1 in the right-hand factor and add these products to get the result. It is important to note that a requirement for matrix multiplication is that the number of columns in the left-hand factor must equal the number of rows in the right-hand factor.

Solve Two Variable Linear Systems Using Cramer's Rule and Inverse Matrices

While there are many applications involving the use of matrices, we will only consider the application of solving systems of equations. Consider the system of equations:

$$5x - 6y = 148$$
$$3x + 2y = 16$$

Written as a matrix equation, this would read

$$\begin{bmatrix} 5 & -6 \\ 3 & 2 \end{bmatrix} \begin{bmatrix} x \\ y \end{bmatrix} = \begin{bmatrix} 148 \\ 16 \end{bmatrix}$$

Notice that the first matrix on the left is the coefficient matrix and the second matrix is the variable matrix. The coefficient matrix has two rows and two columns, giving it the dimensions 2×2, and the variable matrix has two rows and one column. The dimensions of the variable matrix are 2×1. The constant matrix on the right side of the equation has the same dimensions as the variable matrix.

There is a feature of square matrices called a **determinant,** which is used in the analysis of matrices. We will look at how the value of the determinant is computed and then apply this process to a technique called Cramer's rule that is used to solve systems of equations. Consider this matrix:

$$\begin{bmatrix} a & b \\ c & d \end{bmatrix}$$

The determinant for this matrix is $ad - bc$. The determinant for the coefficient matrix in our example is $10 - (-18) = 28$.

For matrices that have more than two rows and columns, the process is a bit trickier. There is an alternative process that we will use and it is similar to what is done for the 2×2 matrix. Observe that we multiply the elements along the main diagonal (ad) and the elements along the minor diagonal (bc) and subtract the result. We apply a variation of this process for the matrices with larger dimensions.

Let's work with this matrix:

$$\begin{bmatrix} 2 & 4 & -1 \\ 5 & -3 & 2 \\ 8 & 1 & 2 \end{bmatrix}$$

We'll copy the matrix, without the brackets, and we will repeat the first two columns of the matrix to the right of the matrix. Why? Because it is a convenient approach to use that is equivalent to the more formal approach of computing determinants for matrices.

$$
\begin{array}{ccccc}
2 & 4 & -1 & 2 & 4 \\
5 & -3 & 2 & 5 & -3 \\
8 & 1 & 2 & 8 & 1
\end{array}
$$

Compute the products on the major (down) diagonals:

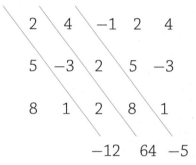

$$
\begin{array}{ccccc}
2 & 4 & -1 & 2 & 4 \\
5 & -3 & 2 & 5 & -3 \\
8 & 1 & 2 & 8 & 1 \\
& & -12 & 64 & -5
\end{array}
$$

Compute the products for the minor (up) diagonals:

$$
\begin{array}{ccccc}
& & 24 & 4 & 40 \\
2 & 4 & -1 & 2 & 4 \\
5 & -3 & 2 & 5 & -3 \\
8 & 1 & 2 & 8 & 1
\end{array}
$$

Take the sum of the major diagonals ($-12 + 64 + -5 = 47$) and subtract the sum of the minor diagonals ($24 + 4 + 40 = 68$): $47 - 68 = -21$. The determinant of this matrix is -21.

▶ Determine the value of the determinant for the following matrix:

$$\begin{bmatrix} 5 & 3 & 2 \\ 8 & -5 & -1 \\ 4 & -1 & 3 \end{bmatrix}$$

▶ Rewrite the matrix, adding the extra columns.

$$\begin{matrix} 5 & 3 & 2 & 5 & 3 \\ 8 & -5 & -1 & 8 & -5 \\ 4 & -1 & 3 & 4 & -1 \end{matrix}$$

▶ Compute the product of the main diagonals.

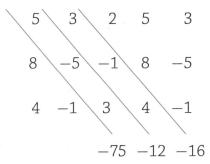

$$-75 \quad -12 \quad -16$$

$$-75 + (-12) + (-16) = -103.$$

▶ Compute the product of the minor diagonals.

$$-40 \quad 5 \quad 72$$

$$\begin{matrix} 5 & 3 & 2 & 5 & 3 \\ 8 & -5 & -1 & 8 & -5 \\ 4 & -1 & 3 & 4 & -1 \end{matrix}$$

$$-40 + 5 + 72 = 37$$

▶ The value of the determinant is $-103 - 37 = -140$.

Unfortunately, the process of using the diagonals does not work for a matrix with dimensions 4×4. As you will see at the end of the section, this does not impact us at this time. (The process for finding the determinant of a 4×4, or larger, matrix uses a process known as minors and is taught in a class on matrix algebra. This is well beyond the scope of pre-calculus.)

Cramer's rule is a technique for solving systems of equations using determinants. You first compute the determinant for the coefficient matrix, D. To determine the value of the first variable (which is traditionally x), replace the first column of the coefficient matrix with the constant matrix. Compute the determinant for this matrix, D_x. The value of x is determined by finding the quotient of these determinants. That is, $x = \dfrac{D_x}{D}$. To find the second variable, replace the second column in the coefficient matrix with the constant matrix. Continue for as many variables as are in the system.

EXAMPLE

▶ Use Cramer's rule to solve the following:

$$5x - 6y = 148$$
$$3x + 2y = 16$$

▶ The determinant for the coefficient matrix is 28. Therefore, $D = 28$.

▶ Replace the first column in the coefficient matrix with the constant matrix.

$$\begin{vmatrix} 148 & -6 \\ 16 & 2 \end{vmatrix}$$

▶ Compute the determinant: $D_x = (148)(2) - (16)(-6) = 296 + 96 = 392$.

▶ Determine the value of x: $x = \dfrac{392}{28} = 14$.

▶ Replace the second column in the coefficient matrix with the constant matrix.

$$\begin{vmatrix} 5 & 148 \\ 3 & 16 \end{vmatrix}$$

▶ Compute the determinant: $D_y = (5)(16) - (3)(148) = 80 - 444 = -364$

▶ Determine the value of y: $y = \dfrac{-364}{28} = -13$

▶ The solution to the problem is the ordered pair (14, −13).

I think we can agree that this is more work than it would be to solve the system by more traditional methods! However, Cramer's rule does prove to be more efficient when the system contains more than two equations in two variables.

▶ Use Cramer's rule to solve the system of equations:

$$5x + 3y + 2z = 12$$
$$8x - 5y - z = 29$$
$$4x - y + 3z = 33$$

▶ The coefficient matrix for this problem is as shown:

$$\begin{bmatrix} 5 & 3 & 2 \\ 8 & -5 & -1 \\ 4 & -1 & 3 \end{bmatrix}$$

▶ The determinant for the coefficient matrix is −140 (we did this problem earlier in this section).

▶ Replace the first column in the coefficient matrix with the constant matrix.

$$\begin{bmatrix} 12 & 3 & 2 \\ 29 & -5 & -1 \\ 33 & -1 & 3 \end{bmatrix}$$

▶ Compute the determinant for this matrix:

$$
\begin{array}{ccccc}
12 & 3 & 2 & 12 & 3 \\
29 & -5 & -1 & 29 & -5 \\
33 & -1 & 3 & 33 & -1 \\
 & -180 & -99 & -58 &
\end{array}
\qquad
\begin{array}{ccccc}
 & & -330 & 12 & 261 \\
12 & 3 & 2 & 12 & 3 \\
29 & -5 & -1 & 29 & -5 \\
33 & -1 & 3 & 33 & -1
\end{array}
$$

▶ The determinant is equal to $(-180 - 99 - 58) - (-330 + 12 + 261)$
$= -280.$

▶ Compute the value of x: $x = \dfrac{-280}{-140} = 2.$

▶ Replace the second column in the coefficient matrix with the constant matrix.

$$
\begin{bmatrix}
5 & 12 & 2 \\
8 & 29 & -1 \\
4 & 33 & 3
\end{bmatrix}
$$

▶ Compute the determinant for the matrix.

$$
\begin{array}{ccccc}
5 & 12 & 2 & 5 & 12 \\
8 & 29 & -1 & 8 & 29 \\
4 & 33 & 3 & 4 & 33 \\
 & 435 & -48 & 528 &
\end{array}
\qquad
\begin{array}{ccccc}
 & & 232 & -165 & 288 \\
5 & 12 & 2 & 5 & 12 \\
8 & 29 & -1 & 8 & 29 \\
4 & 33 & 3 & 4 & 33
\end{array}
$$

▶ The determinant is equal to $(435 - 48 + 528) - (232 - 165 + 288) = 560.$

▶ Compute the value of y: $y = \dfrac{-560}{-140} = -4$

▶ We can repeat this process to find the value of z, but it is a lot easier to substitute the values found for x and y into one of the equations.

$$5x + 3y + 2z = 12$$
$$\Rightarrow 5(2) + 3(-4) + 2z = 12$$
$$\Rightarrow 2z = 14$$
$$\Rightarrow z = 7$$

▶ The solution is the ordered triple (2, −4, 7).

I believe this is quicker than the elimination method usually used to solve a system of three equations in three variables, and I can guarantee you that for systems involving more than three equations in the same number of variables, Cramer's rule is absolutely quicker than the elimination method.

There is a technique that is still quicker than either the elimination method or Cramer's rule and that is the process of using the inverse of the coefficient matrix to solve the system. All systems of equations can be written in the form $AX = B$, where A is the coefficient matrix, X is the matrix containing the variables, and B is the matrix of constants. The solution to this equation is $X = A^{-1}B$. (In a traditional Algebra I problem, the solution to the equation $ax = b$ is $x = \dfrac{b}{a}$. However, there is no such operation as matrix division, so we use the multiplicative inverse of the coefficient matrix to determine the solution.)

All of this work can be done on the calculator. You just need to be careful when you enter the matrices.

EXAMPLE

▶ Solve:

$$5x - 6y = 148$$
$$3x + 2y = 16$$

▶ The matrix equation for this problem is

$$\begin{bmatrix} 5 & -6 \\ 3 & 2 \end{bmatrix}\begin{bmatrix} x \\ y \end{bmatrix} = \begin{bmatrix} 148 \\ 16 \end{bmatrix}$$

▶ The solution to this system is

$$\begin{bmatrix} x \\ y \end{bmatrix} = \begin{bmatrix} 5 & -6 \\ 3 & 2 \end{bmatrix}^{-1}\begin{bmatrix} 148 \\ 16 \end{bmatrix} = \begin{bmatrix} 14 \\ -13 \end{bmatrix}$$

▶ We now know that the solution is the ordered pair (14, −13).

Apply the inverse matrix approach to solving systems of equations to a system with three equations and three variables.

EXAMPLE

▶ Solve:

$$5x + 3y + 2z = 12$$
$$8x - 5y - z = 29$$
$$4x - y + 3z = 33$$

▶ The matrix equation for this problem is

$$\begin{bmatrix} 5 & 3 & 2 \\ 8 & -5 & -1 \\ 4 & -1 & 3 \end{bmatrix}\begin{bmatrix} x \\ y \\ z \end{bmatrix} = \begin{bmatrix} 12 \\ 29 \\ 33 \end{bmatrix}$$

▶ The solution to the system is

$$\begin{bmatrix} x \\ y \\ z \end{bmatrix} = \begin{bmatrix} 5 & 3 & 2 \\ 8 & -5 & -1 \\ 4 & -1 & 3 \end{bmatrix}^{-1}\begin{bmatrix} 12 \\ 29 \\ 33 \end{bmatrix} = \begin{bmatrix} 2 \\ -4 \\ 7 \end{bmatrix}$$

▶ The solution is the ordered triple (2, −4, 7).

You have to agree this is a much faster approach to solving systems of linear equations!

Solving Three or More Variable Linear Systems

Let's apply what we've learned about solving systems of equations. You will find that the emphasis will now be on determining the definition of the variables and writing the correct equations and less on the process of solving the system of equations.

▶ Joanne is playing a word game similar to one that is currently a popular application. The goal is to use the letters on her rack to get as many points as she can. There are four squares available to her:

Triple Letter		Double Letter	

▶ Joanne has the letters A, E, L, and P to use. If she plays *PLEA*, she gets 27 points. If she plays *PALE*, she gets 26 points. She scores 22 points for *LEAP* and 28 points for *PEAL*. How many points is each letter worth?

▶ We can easily define the variables as A, E, L, and P at the point value for each letter. The resulting equations are

$$PLEA : 3P + L + 2E + A = 27$$
$$PALE : 3P + A + 2L + E = 26$$
$$LEAP : 3L + E + 2A + P = 22$$
$$PEAL : 3P + E + 2A + L = 28$$

▶ We should rewrite the equations so that the like variables are in the same column.

$$PLEA : 3P + L + 2E + A = 27$$
$$PALE : 3P + 2L + E + A = 26$$
$$LEAP :\ P + 3L + E + 2A = 22$$
$$PEAL : 3P + L + E + 2A = 28$$

▶ The matrix equation for this system is

$$\begin{bmatrix} 3 & 1 & 2 & 1 \\ 3 & 2 & 1 & 1 \\ 1 & 3 & 1 & 2 \\ 3 & 1 & 1 & 2 \end{bmatrix} \begin{bmatrix} P \\ L \\ E \\ A \end{bmatrix} = \begin{bmatrix} 27 \\ 26 \\ 22 \\ 28 \end{bmatrix}$$

▶ The solution to this system is

$$\begin{bmatrix} P \\ L \\ E \\ A \end{bmatrix} = \begin{bmatrix} 3 & 1 & 2 & 1 \\ 3 & 2 & 1 & 1 \\ 1 & 3 & 1 & 2 \\ 3 & 1 & 1 & 2 \end{bmatrix}^{-1} \begin{bmatrix} 27 \\ 26 \\ 22 \\ 28 \end{bmatrix} = \begin{bmatrix} 5 \\ 2 \\ 3 \\ 4 \end{bmatrix}$$

▶ The P is worth 5 points, the L is worth 2 points, the E is worth 3 points, and the A is worth 4 points.

Production costs for a manufacturing concern are not necessarily consistent from day to day.

EXAMPLE

▶ Martin was looking over the ledgers for his business and noted the production costs for a five-week period. In week 1, the company produced 8 units on Monday, 12 on Tuesday, 11 on Wednesday, 9 on

Thursday, and 10 on Friday. The production for the next four weeks was as follows:

Week 2: 11 Monday, 12 Tuesday, 15 Wednesday, 10 Thursday, 5 Friday
Week 3: 7 Monday, 15 Tuesday, 20 Wednesday, 13 Thursday, 11 Friday
Week 4: 17 Monday, 20 Tuesday, 18 Wednesday, 7 Thursday, 1 Friday
Week 5: 15 Monday, 18 Tuesday, 21 Wednesday, 8 Thursday, 3 Friday

▶ The production costs for week 1 were $4,115, $4,255 for week 2, $5,510 for week 3, $4,820 for week 4, and $5,105 for week 5. What is the production cost for each day of the week?

▶ Let M represent the production cost for Monday, T represent the production cost for Tuesday, W represent the production cost for Wednesday, H represent the production cost for Thursday, and F represent the production cost for Friday.

▶ The system of equations is

$$\begin{aligned}
8M + 12T + 11W + 9H + 10F &= 4115 \\
11M + 12T + 15W + 10H + 5F &= 4255 \\
7M + 15T + 20W + 13H + 11F &= 5510 \\
17M + 20T + 18W + 7H + 1F &= 4820 \\
15M + 18T + 21W + 8H + 3F &= 5105
\end{aligned}$$

▶ The variables are aligned in appropriate columns so we can write the matrix equation.

$$\begin{bmatrix} 8 & 12 & 11 & 9 & 10 \\ 11 & 12 & 15 & 10 & 5 \\ 7 & 15 & 20 & 13 & 11 \\ 17 & 20 & 18 & 7 & 1 \\ 15 & 18 & 21 & 8 & 3 \end{bmatrix} \begin{bmatrix} M \\ T \\ W \\ H \\ F \end{bmatrix} = \begin{bmatrix} 4115 \\ 4225 \\ 5510 \\ 4820 \\ 5105 \end{bmatrix}$$

▶ The solution to this system is

$$
\begin{bmatrix} M \\ T \\ W \\ H \\ F \end{bmatrix} = \begin{bmatrix} 8 & 12 & 11 & 9 & 10 \\ 11 & 12 & 15 & 10 & 5 \\ 7 & 15 & 20 & 13 & 11 \\ 17 & 20 & 18 & 7 & 1 \\ 15 & 18 & 21 & 8 & 3 \end{bmatrix}^{-1} \begin{bmatrix} 4{,}115 \\ 4{,}225 \\ 5{,}510 \\ 4{,}820 \\ 5{,}105 \end{bmatrix} = \begin{bmatrix} 65 \\ 70 \\ 90 \\ 85 \\ 100 \end{bmatrix}
$$

▶ Production costs are $65 per unit on Monday, $70 per unit on
Tuesday, $90 per unit on Wednesday, $85 per unit on Thursday, and
$100 per unit on Friday.

Solving Systems of Linear Inequalities

You've solved systems of linear inequalities in Algebra I and Algebra II.
In this section we look at an application of these systems called **linear
programming**. We'll start with a basic problem, discuss the process for
answering the problem, and then look at applications.

   With linear programming programs, there is an **objective function**. This
is the item whose "best" value we are trying to find. The "best" value could
be a maximum or minimum value. There is a set of constraints that the
problem must satisfy. Constraints are written in the form of inequalities.
The first step in the solution is to graph all the constraints and find the
region that satisfies all the constraints. This region is called the **feasibility
region**. The last step, which is based on mathematical theory that you may
someday study in college, is to evaluate the objective function with the
coordinates of the corner points of the feasibility region, because the theory
upon which linear programming is based tells us that is where the optimum
solution is located:

▶ Determine the maximum value of $f = 5x + 7y$ subject to the following constraints:

$$x \geq 0, \; y \geq 0$$
$$x + 2y \leq 12$$
$$2x + y \leq 12$$

▶ Step 1. Graph the constraints to determine the feasibility region. Notice that the region resides in the first quadrant.

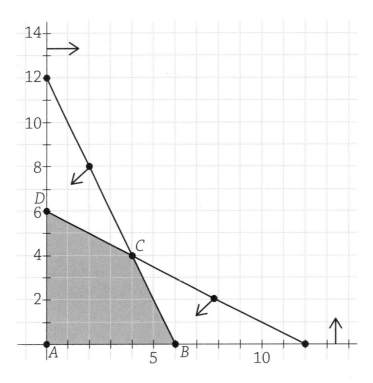

The arrows indicate the direction for the shading for each inequality and, for the ease of reading, only the common region is displayed. The corner points for the feasibility are labeled.

▶ Step 2. Determine the coordinate of the corner points for the feasibility region.

▶ The coordinates of these points are $A(0,0), B(6,0), C(4,4),$ and $D(0,6)$.

▶ Step 3. Evaluate the feasibility function with the coordinates of the corner points.

▶ The objective function is evaluated with each of these points:

$$f(A) = 5(0) + 7(0) = 0$$
$$f(B) = 5(6) + 7(0) = 30$$
$$f(C) = 5(4) + 7(4) = 48$$
$$f(D) = 5(0) + 7(6) = 42$$

▶ Step 4. Choose the optimum solution.

▶ The maximum value of the objective function is 48.

There are times when the feasibility region will be unbounded. Usually these problems seek the minimum value for the objective function.

EXAMPLE

▶ Determine the minimum value of $f = 5x + 7y$ subject to the following constraints:

$$x \geq 0, y \geq 0$$
$$x + 2y \geq 12$$
$$2x + y \geq 12$$

▶ Step 1. Determine the feasibility region.

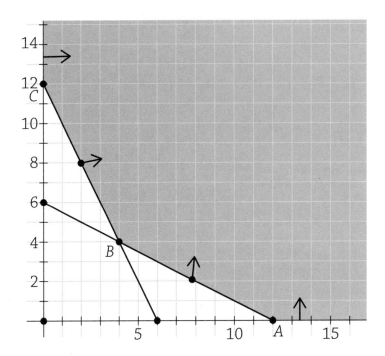

▶ Step 2. The coordinates of the three corner points are $A(12, 0)$, $B(4, 4)$, and $C(0, 12)$.

▶ Step 3. Evaluate the objective function.

$$f(A) = 5(12) + 7(0) = 60$$
$$f(B) = 5(4) + 7(4) = 48$$
$$f(C) = 5(0) + 7(12) = 84$$

▶ Step 4. The minimum value of the objective function is 48.

It's time to take a look at an application.

▶ Ma and Pa Kettle produce 4-quart and 6-quart pots. Each 4-quart pot requires 100 square inches of aluminum and takes 12 minutes to mold and refine into a finished pot. Each 6-quart pot requires 150 square inches of aluminum and takes 15 minutes to mold and refine into a finished pot. The Kettles have a tradition that they make at least 20 four-quart pots and 10 six-quart pots each day.

▶ The Kettle production process has a daily limitation of 11,000 square inches of aluminum and 20 hours of labor. If the profit for a 4-quart pot is $4.50 and the profit for a 6-quart pot is $5.25, how many pots of each kind should the Kettles produce in order to maximize their daily profit? (With thanks to Marjorie Main and Percy Kilbride for the laughs they provided.)

▶ The objective function is usually pretty easy to identify—look for the quantity that needs to be the largest or smallest. In this case the profit is to be maximized. What is being sold? The Kettles are selling 4-quart and 6-quart pots. We have just identified our variables.

▶ Let x represent the number of 4-quart pots made, and let y represent the number of 6-quart pots made.

▶ Object function: $P(x, y) = 4.50x + 5.25y$

▶ What are the constraints on the production process? First, there is the physical reality that a negative number of pots cannot be produced, so $x \geq 0, x \geq 0$.

▶ The Kettles have a tradition for a minimum number of each pot made daily: $x \geq 20, y \geq 10$.

▶ They have a limited amount of aluminum that they can use each day: $100x + 150y \leq 11,000$.

▶ They have a limited labor supply. Be careful to note that the time is allocated in minutes but the amount of time available is measured in hours. Be sure that the time units agree when writing the constraint: $12x + 15y \leq 1200$.

▶ We are now ready to determine the feasibility region.

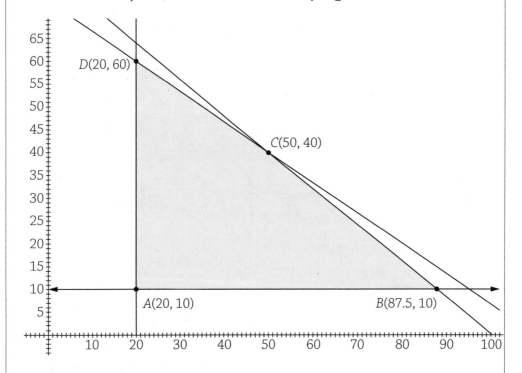

▶ The vertices of the corner of the feasibility region are $A(20, 10)$, $B(87, 5, 10)$, $C(50, 40)$, and $D(20, 60)$.

▶ The value of the objective function for each of these points is

$$P(A) = 4.5(20) + 5.25(10) = 142.50$$
$$P(B) = 4.5(87.5) + 5.25(10) = 446.25$$
$$P(A) = 4.5(50) + 5.25(40) = 435$$
$$P(A) = 4.5(20) + 5.25(60) = 405$$

▶ The mathematical model indicates that the maximum profit will occur at point *B*. Make 87.5 four-quart pans and 10 six-quart pans for a profit of $446.25. Wait, you can't make half a pot! True. So, you make 87 four-quart pots and 10 six-quart pots daily for a maximum daily profit of $444.

That was a good example of how the mathematics needs to be adjusted to meet reality. Here is another application of linear programming.

▶ Daniel and his nutritionist worked out a dietary plan that would require Daniel to get at least 150 units of vitamin B9, at least 19 units of vitamin D, and at least 720 units of vitamin C daily. To achieve this, the nutritionist recommended two vitamin supplements, Full Day and Good4U.

▶ Each dose of Full Day contains 6 units of vitamin B9, 0.1 units of vitamin D, and 6 units of vitamin C. One dose of Full Day costs $0.025. Each dosage of Good4U contains 1 unit of vitamin B9, 0.2 units of vitamin D, and 5 units of vitamin C. One dose of Good4U costs $0.02. How many units of each supplement should Daniel take each day to meet his dietary needs at a minimum cost?

▶ The goal here is to achieve a minimum cost. Daniel is spending money for each dose of Full Day and Good4U that he buys. Therefore, let x represent the number of doses of Full Day taken daily and y represent the number of doses of Good4U taken daily.

▶ Constraints: Daniel cannot take a negative number of doses of either supplement, so $x \geq 0$ and $y \geq 0$.

▶ The three vitamins that Daniel's plan works to increase are vitamins B9, D, and C. The doses he takes must meet these requirements:

$$B9: 6x + y \geq 150$$
$$D: 0.01x + 0.2y \geq 19$$
$$C: 6x + 5y \geq 720$$

▶ The cost function for these supplements will be $C(x, y) = 0.025x + 0.02y$.

▶ The feasibility region is

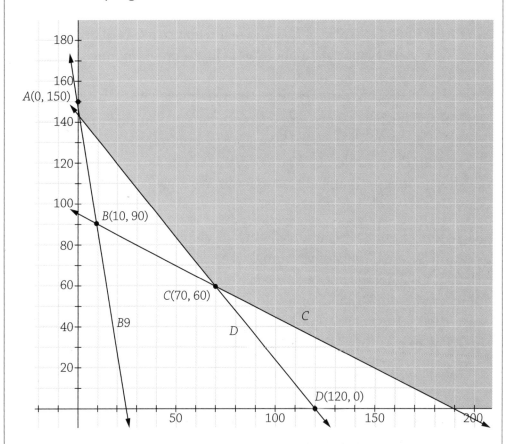

▶ The corner points for the feasibility region have coordinates $A(0, 150)$, $B(10, 90)$, $C(70, 60)$, and $D(120, 0)$. The value of the objective function at each of these points is

$$C(A) = 0.025(0) + 0.02(150) = 3.00$$
$$C(B) = 0.025(10) + 0.02(90) = 2.05$$
$$C(C) = 0.025(70) + 0.02(60) = 2.95$$
$$C(D) = 0.025(120) + 0.02(0) = 3.00$$

▶ Daniel should take 10 doses of Full Day and 90 doses of Good4U to meet his requirements at a cost of $2.05 per day.

A question that must be considered is this: What happens if there are more than two variables that must be considered in a linear programming scenario? It turns out that there is a very elegant matrix solution that can be used to solve such problems. The setup and solution are fairly complicated, and we'll leave that for your future study (if you are so inclined)!

EXERCISES

EXERCISE 8-1

Follow the instructions for questions 1 and 2.

1. If $\begin{bmatrix} 4x + 3y \\ 5x - 2y \end{bmatrix} = \begin{bmatrix} 43 \\ 25 \end{bmatrix}$, determine the values of x and y.

2. Compute $\begin{bmatrix} 4 & 9 & -3 \\ 2 & -5 & 1 \end{bmatrix} \begin{bmatrix} 10 \\ 2 \\ 8 \end{bmatrix}$

Use Cramer's rule to solve questions 3 and 4.

3. $3x + 2y = 6$
    $4x - 5y = 77$

4. $5x + 3y - 6z = 11$
    $8x - 2y + 5z = 43$
    $2x + 3y + 4z = 49$

Use matrix equations to solve questions 5 to 8.

5. $3x + 2y = 6$
    $4x - 5y = 77$

6. $5x + 3y - 6z = 11$
    $8x - 2y + 5z = 43$
    $2x + 3y + 4z = 49$

7. $4w - 6x + 3y + 2z = 48$
    $8w + 2x + 5y - 7z = -116$
    $-5w - 8x - 4y + 8z = 141$
    $20w + 13x + 19y + 25z = 245$

8. $8w + 9x - 12y + 12z = 13$
   $6w - 3x + 4y + 6z = -9$
   $10w + 9x - 4y - 6z = -11$
   $2w + 15x + 8y + 18z = 16$

Answer questions 9 and 10.

9. The points (3, 7), (−2, −33), and (−5, −153) lie on the parabola with equation $y = ax^2 + bx + c$. Find the coordinates of the vertex of the parabola.

10. Eileen is playing a word game similar to one that is currently a popular application. The goal is to use the letters on her rack to get as many points as she can. There are four squares available to her:

Triple Letter		Double Letter	

Eileen has the letters *A, E, L,* and *V* to use. If she plays *VALE*, she gets 25 points. If she plays *VEAL*, she gets 26 points. She scores 20 points for *LAVE* and 20 points for *EVAL*. How many points is each letter worth?

EXERCISE 8-2

Determine the optimum solution for each of the linear programming problems in questions 1 to 3.

1. Maximize $f(x, y) = 8x + 7y$ subject to
   $x \geq 0 \quad y \geq 0$
   $x \geq 3 \quad y \geq 5$
   $5x + 6y \leq 120$

2. Maximize $g(x, y) = 80x + 75y$ subject to
   $x \geq 0 \quad y \geq 0$
   $y \leq 2x$
   $4x + y \leq 18$
   $x + y \geq 6$

3. Minimize $q(x, y) = 30x + 45y$ subject to
   $$x \geq 0 \quad y \geq 0$$
   $$x + y \geq 10$$
   $$2x + y \geq 12$$
   $$x + 2y \geq 12$$

Answer each of the problems in questions 4 and 5.

4. Patty's Burgers make their burgers from ground beef and ground pork. Patty's Pig uses 4 ounces of pork and 4 ounces of beef, while Patty's What's Your Beef burger uses 2 ounces of pork and 4 ounces of beef. Patty's has 600 ounces of ground pork and 720 ounces of ground beef to work with each day. Each Patty's Pig sandwich costs $6, while each What's Your Beef sandwich costs $5. If the number of Patty's Pig sandwiches is always, at most, twice the number of the What's Your Beef sandwiches, how many of each sandwich should they make in order to maximize their revenue?

5. Kirsten volunteers to bring origami dragons and butterflies to sell at a crafts fair. It takes her 4 minutes to make a butterfly and 6 minutes to make a dragon. She plans to sell the butterflies for $5 each and the dragons for $6 each. She does not want to make any more than 15 butterflies. If she has only 24 pieces of origami paper and can't spend more than 2 hours folding, how many of each animal should Kirsten make to maximize her profit?

Flashcard App

 **Triangle Trigonometry**

MUST KNOW

⚡ Trigonometry is based on the relationships among the sides and angles of a right triangle.

⚡ The Law of Sines and the Law of Cosines enable us to solve problems in triangles without a right angle.

⚡ The Law of Sines enables us to use the relationship among two sets of angles and the opposite-side lengths to solve for a missing measurement, given the values of three of the four angles and sides.

⚡ The Law of Cosines lets us solve triangle-related problems if we have the values of the lengths of the three sides or the values of two sides and their included angle available.

rigonometry started as an application of similar right triangles. It then evolved into an application for any triangle. When these triangles were placed on a circle, the circular functions were developed and with them applications of period phenomena were modeled with trigonometric functions.

Right Triangle Trigonometry

We can remember the three basic functions—sine, cosine, and tangent—using the mnemonic SOHCAHTOA. The sine ratio (S) is the ratio of the lengths of the side opposite (O) to the acute angle to the hypotenuse (H), the cosine (C) is the ratio of the adjacent (A) side to the hypotenuse (H), and the tangent (T) is the ratio of the opposite (O) side to the adjacent (A) side. Each ratio is abbreviated with three letters. The sine ratio is abbreviated as sin, the cosine ratio as cos, and the tangent ratio as tan.

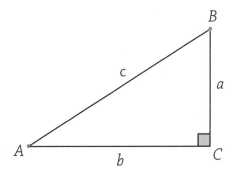

In the diagram above, angle C is a right angle. The sine of angle A is written $\sin(A) = \frac{a}{c}$; the cosine of angle A, $\cos(A) = \frac{b}{c}$; and the tangent of angle A, $\tan(A) = \frac{a}{b}$. The ratios for angle B are $\sin(B) = \frac{b}{c}$, $\cos(B) = \frac{a}{c}$, and $\tan(B) = \frac{b}{a}$.

Angles A and B are complementary angles. In fact, the word cosine comes from complement of the sine. For any acute angle, it will always be the case that $\sin(A) = \cos(90 - A)$.

EXAMPLE

▶ Given the right triangle in the accompanying diagram, find, to the nearest tenth, IJ.

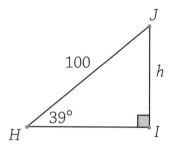

▶ IJ is opposite the angle and the hypotenuse is 100. Therefore, $\sin(39) = \dfrac{h}{100}$ so that $h = 100\sin(39)$.

▶ With your calculator set in degree mode, compute $100\sin(39)$ to equal 62.9.

Let's take a look at another problem.

EXAMPLE

▶ Given the right triangle in the accompanying diagram, find, to the nearest tenth, KL.

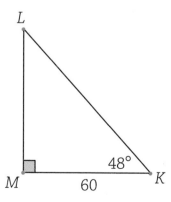

▶ 60 is the length of the adjacent side and *KL* is the length of the hypotenuse. Use the cosine ratio to write $\cos(48) = \dfrac{60}{KL}$. Multiply to get $KL\cos(48) = 60$ so that $KL = \dfrac{60}{\cos(48)}$.

▶ Use your calculator to determine $KL = 89.7$.

Here's a problem that can be solved in two ways.

▶ Find the length of *GH* in the accompanying diagram. Round your answer to the nearest tenth.

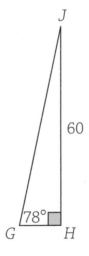

▶ *JH* is opposite the angle and *GH* is adjacent to the angle. Use the tangent ratio to write $\tan(78) = \dfrac{60}{GH}$. Solve for *GH*: $GH = \dfrac{60}{\tan(78)} = 12.8$.

▶ You should notice that the measure of angle *J* is 12°. This problem could also be solved using the equation $\tan(12) = \dfrac{GH}{60}$ so that $GH = 60\tan(12)$.

In this next example we are given the lengths of two sides and asked to find the measure of the angle.

EXAMPLE

▶ Find, to the nearest degree, the measure of θ in the accompanying diagram.

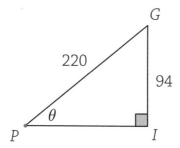

θ is the Greek letter theta and is often used in mathematics to represent angles.

▶ The two sides whose lengths are given in the problem are the hypotenuse and the side opposite θ. Use the sine ratio to write $\sin(\theta) = \dfrac{94}{220}$.

▶ The inverse sine function, $\sin^{-1}$, on your calculator gives the measure of the acute angle whose sine is entered. Enter $\sin^{-1}\left(\dfrac{94}{220}\right)$ to get $\theta = 25°$.

The next example requires the use of two right triangles to solve the problem.

EXAMPLE

▶ Mark and Sally had always wondered how tall Wagner Hill was. After a lesson in right triangle trigonometry, they decided to finally get the answer.

▶ They used an inclinometer—an actual word and device!—and found the angle of elevation to the top of Wagner Hill was 23°. They walked 500 feet directly toward Wagner Hill and took a second measurement. This time the angle of elevation was 38°. They went back home to work with their data.

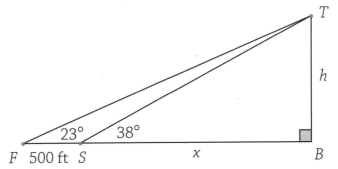

▶ They let x represent the distance from S, the point of their second reading, to the foot of the hill, B. They were able to write the equation $\tan(23°) = \dfrac{h}{500 + x}$. Solving for h, they wrote $h = (500 + x)\tan(23°) = 500\tan(23°) + x\tan(23°)$. Solving for x, they had the equation $x = \dfrac{h - 500\tan(23°)}{\tan(23°)}$.

▶ Working with $\triangle FBT$, they wrote $\tan(38°) = \dfrac{h}{x}$ or $x = \dfrac{h}{\tan(38°)}$.

Substituting for x in the first equation, Mark and Sally finally had an equation they could solve, but not without first remarking on how ugly the equation was!

$$\frac{h}{\tan(38°)} = \frac{h - 500\tan(23°)}{\tan(23°)}$$

$$h\tan(23°) = (h - 500\tan(23°))\tan(38°)$$

$$h\tan(23°) = h\tan(38°) - 500\tan(23°)\tan(38°)$$

$$h\tan(23°) - h\tan(38°) = -500\tan(23°)\tan(38°)$$

$$h(\tan(23°) - \tan(38°)) = -500\tan(23°)\tan(38°)$$

$$h = \frac{-500\tan(23°)\tan(38°)}{\tan(23°) - \tan(38°)} = \frac{500\tan(23°)\tan(38°)}{\tan(38°) - \tan(23°)} = 464.7$$

▶ Mark and Sally were able to report to their class that the height of Wagner Hill is approximately 465 feet.

▶ We will look at this problem again in a later section of the chapter. You'll find that solution takes fewer steps.

Before we move on to new topics, there are two special right triangles that are often used in trigonometry: the **isosceles right triangle** and the **30-60-90 triangle**.

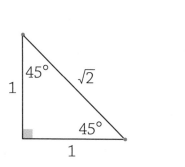 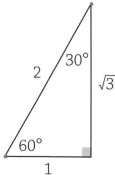

The values of the trigonometric functions for these angles are worth memorizing.

	30°	45°	60°
sine	$\dfrac{1}{2}$	$\dfrac{\sqrt{2}}{2}$	$\dfrac{\sqrt{3}}{2}$
cosine	$\dfrac{\sqrt{3}}{2}$	$\dfrac{\sqrt{2}}{2}$	$\dfrac{1}{2}$
tangent	$\dfrac{\sqrt{3}}{3}$	1	$\sqrt{3}$

Figuring Out the Area of a Triangle Using Trigonometry

The area of $\triangle STU$, below, is $\frac{1}{2}(ST)(UV)$, with UV being the length of the altitude to $\overline{ST}$.

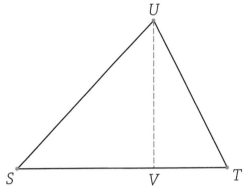

In right $\triangle STU$, $\sin(S) = \dfrac{UV}{SV}$ so that $UV = SV\sin(S)$. Substitute this into the formula for the area of the triangle and get the result that the area of $\triangle STU$ is $\frac{1}{2}(ST)(SV)\sin(S)$.

In general, the area of a triangle is equal to one-half the product of the lengths of two sides of a triangle and the sine of the included angle.

▶ Given $\triangle ABC$ with $AB = 40$ cm, $AC = 48$ cm, and m$\angle A = 67°$, find the area of the triangle to the nearest tenth of a square centimeter.

▶ The area of $\triangle ABC = \frac{1}{2}(40)(48)\sin(67) = 883.7$ sq cm.

How is this formula used if the angle in question is obtuse?

EXAMPLE

▶ Given $\triangle ABC$ with $AB = 40$ cm, $AC = 48$ cm, and m$\angle A = 113°$, find the area of the triangle to the nearest tenth of a square centimeter.

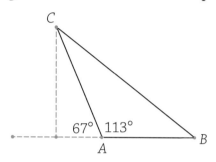

▶ The altitude is exactly the same for this triangle as in the last problem.

▶ The area of $\triangle ABC$ is $\frac{1}{2}(40)(48)\sin(113) = 883.7$ sq. cm.

▶ This example is a reminder that the sine of the supplement of an acute angle is equal to the sine of the angle, i.e., $\sin(180 - A) = \sin(A)$.

We need to be careful when we are asked to find the measure of the included angle when given the lengths of two sides of the triangle and the area of the triangle.

EXAMPLE

▶ Given $\triangle QRS$ has an area of 480 square inches, if $QR = 60$ in and $RS = 50$ in, find, to the nearest tenth of a degree, the measure of angle R.

▶ The area of $\triangle QRS$ is $\frac{1}{2}(QR)(RS)\sin(R)$, so $480 = \frac{1}{2}(60)(50)\sin(R)$. Solve this equation to get $\sin(R) = \dfrac{480}{1500}$.

▶ If $\sin(R) = \dfrac{480}{1500}$, then $R = \sin^{-1}\left(\dfrac{480}{1500}\right) = 18.7°$. However, angle R could be an obtuse angle.

▶ Therefore, m$\angle R = 18.7°$ or $161.3°$ (the supplement of the acute angle).

Law of Sines

The area of $\triangle STU$ is $\frac{1}{2}(ST)(SV)\sin(S)$. It can also be computed with the product $\frac{1}{2}(ST)(TV)\sin(T)$ or $\frac{1}{2}(TV)(SV)\sin(V)$. Since all three expressions represent the area of the same triangle, they must be equal.

$$\frac{1}{2}(ST)(SV)\sin(S) = \frac{1}{2}(ST)(TV)\sin(T) = \frac{1}{2}(TV)(SV)\sin(V)$$

Divide all three expressions by $\frac{1}{2}(ST)(TV)(VS)$ to get the equation that is called the **Law of Sines**:

$$\frac{\sin(V)}{ST} = \frac{\sin(T)}{SV} = \frac{\sin(S)}{VT}$$

Using the convention of naming the side of a triangle with the lowercase letter matching the vertex of the angle, this becomes the more familiar form of the equation $\frac{\sin(V)}{v} = \frac{\sin(T)}{t} = \frac{\sin(S)}{s}$.

It is important to note that the proportion created by any two of the fractions in the Law of Sines involves two sides and two angles and that the orientation of these sides and angles fit the patterns **AAS**, **ASA**, or **SSA** (which you may remember from your study of geometry).

EXAMPLE

In $\triangle ABC$, $BC = 120$ cm, $m\angle A = 56°$, and $m\angle B = 49°$. Find, to the nearest tenth of a centimeter, the lengths of $\overline{AB}$ and $\overline{AC}$.

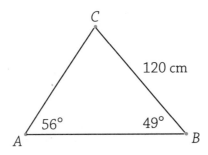

▶ Subtract the sum of 56° and 49° from 180° to determine the measure of the third angle of the triangle, $m\angle C = 75°$ $m\angle C = 75°$. Use the Law of Sines to solve for each side. $\dfrac{\sin(56)}{120} = \dfrac{\sin(49)}{AC}$. Multiply. $AC\sin(56) =$ $120\sin(49)$, so $AC = \dfrac{120\sin(49)}{\sin(56)} = 109.2$.

▶ $\dfrac{\sin(56)}{120} = \dfrac{\sin(75)}{AB}$ becomes $AB\sin(56) = 120\sin(75)$, or

$AB = \dfrac{120\sin(75)}{\sin(56)} = 139.8$

Here is another example of the Law of Sines but with a slight twist.

EXAMPLE

▶ In $\triangle ABC$, $AB = 120$ cm, $m\angle A = 67°$, and $m\angle B = 53°$. Find, to the nearest tenth of an inch, the lengths of $\overline{BC}$ and $\overline{AC}$.

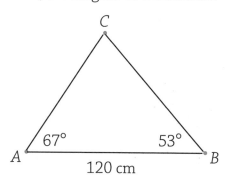

▶ Subtract the sum of 67° and 53° from 180° to determine the measure of the third angle of the triangle, $m\angle B = 60°$. Use the Law of Sines to solve for each side. $\dfrac{\sin(60)}{120} = \dfrac{\sin(53)}{AC}$.

▶ Multiply. $AC\sin(60) = 120\sin(53)$, so $AC = \dfrac{120\sin(53)}{\sin(60)} = 110.7$ cm.

$$\frac{\sin(60)}{120} = \frac{\sin(67)}{BC} \text{ becomes } BC\sin(60) = 120\sin(67),$$

$$\text{so } BC = \frac{120\sin(67)}{\sin(60)} = 127.5 \text{ cm.}$$

What if we are given the lengths of two sides of a triangle and the measure of an angle that is not between these two sides?

EXAMPLE

In $\triangle ABC$, $AB = 80$ cm, $AC = 60$ cm, and m$\angle C = 40°$. Find, to the nearest tenth of a degree, the measure of $\angle A$.

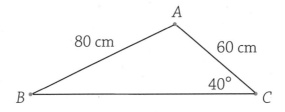

Using $\dfrac{\sin(B)}{b} = \dfrac{\sin(C)}{c}$, the equation becomes $\dfrac{\sin(B)}{60} = \dfrac{\sin(40)}{80}$.

There is a rule from geometry that we must keep in mind when we are trying to find an unknown angle using the Law of Sines: In a triangle, the larger angle is always opposite the large side. In this case, we are being told that m$\angle B = 40°$.

Working this equation we get $\sin(B) = \dfrac{60\sin(40)}{80}$, so that

$$m\angle B = \sin^{-1}\left(\frac{60\sin(40)}{80}\right) = 28.82°.$$

Therefore, the measure for $\angle A$ is $180° - (40° + 28.82°) = 111.18°$, or $111.2°$, rounded to the nearest tenth.

Let's take another look at how Mark and Sally could determine the height of Wagner Hill.

EXAMPLE

▶ To recap: They used an inclinometer and found the angle of elevation to the top of Wagner Hill was 23°. They walked 500 feet directly toward Wagner Hill and took a second measurement. This time the angle of elevation was 38°. They went back home to work with their data.

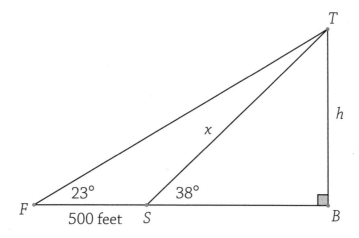

▶ Notice that our figure is a little different this time around. The measure of $\angle FTS$ is 15°. Use the Law of Sines in $\triangle FTS$ to find an expression for x.

$$\frac{x}{\sin(23)} = \frac{500}{\sin(15)}$$

▶ We can write $x = \dfrac{500\sin(23)}{\sin(15)}$.

▶ Now we can use the right triangle: $\sin(38) = \dfrac{h}{x}$, so that $h = x\sin(38)$

▶ Substitute for x: $h = \dfrac{500\sin(23)}{\sin(15)}\sin(38) = 464.7$

▶ The height of Wagner Hill is approximately 465 feet.

The Ambiguous Case

A warning was given in the solution to the last example from the section on the Law of Sines. If the measure of the unknown angle must be less than the measure of the known angle, we can confidently state the measure of the unknown angle after the computations for the Law of Sines are completed. However, if the measure of the unknown angle is greater than the measure of the known angle, we must acknowledge that there are two possible angles whose sine is computed with the Law of Sines. One of those angles is acute (and will be the answer displayed by the calculator) and the other will be obtuse. Fortunately, we know the obtuse angle will be the supplement of the computed acute angle.

Let's take a few moments to examine this statement more closely. Given $\angle A$ and $AR = 30$ cm with m$\angle A = 40°$, what is the length of the shortest segment that can be drawn from point R to create a triangle?

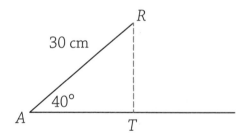

If arcs are drawn from vertex R, as shown in the accompanying diagram, there is one arc that intersects the ray of angle A only once.

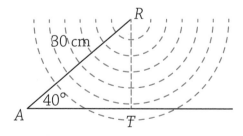

Here are some guidelines for us to consider:

- The point of intersection for this arc is the foot of the perpendicular from R to the ray. (Hopefully this makes sense to you because the shortest distance from a point to a line is along the perpendicular. This is why you are told to stand up "straight" when someone is measuring your height.) If point T is the point at which the perpendicular intersects the side of the ray, then $\triangle ART$ is a right triangle and $AR = 30 \sin(40)$.

- If an arc is drawn with a length RT that is less than this $AR \sin(A)$, the arc will never intersect the side of the ray. It will not be possible to create such a triangle.

- If an arc is drawn with a length RT that is greater than AR, the arc will intersect the ray somewhere to the right and one triangle can be constructed.

- If an arc is drawn with a length RT that is greater than $AR \sin(A)$ but less than AR (i.e., $AR \sin(A) < RT < AR$), then there will be two triangles that can be constructed. This scenario is referred to as the **ambiguous case**.

The situations described above apply when the angle in question is an acute angle. If the given angle is either a right angle or an obtuse angle, the length of the arc drawn from the endpoint of the segment must be longer than the segment itself or the arc will not intersect the other side of angle C.

In $\triangle ART$, $AR = 60$ in, $RT = 30\sqrt{2}$ in, and $m\angle A = 45°$. How many triangles, if any, can be constructed?

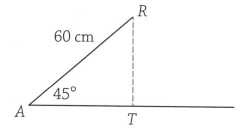

▶ The shortest segment that can be drawn from R to make the triangle is along the perpendicular. The length of the perpendicular is

$60\sin(45) = 60\left(\dfrac{\sqrt{2}}{2}\right) = 30\sqrt{2}$. Therefore, we know that $\triangle ART$ is a right triangle.

▶ There is 1 triangle that can be constructed.

Here is a slightly different version of that problem.

EXAMPLE

▶ In $\triangle ART$, $AR = 60$ in, $RT = 35$ in, and $m\angle A = 45°$. How many triangles, if any, can be constructed?

▶ $60\sin(45) = 42.426$ is the smallest segment that can be constructed to reach the side of the angle. With $RT = 35$, the arc will not intersect the side of the angle, so 0 triangles can be constructed.

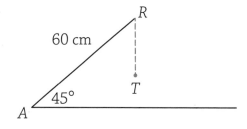

Here is an example where it pays to look at the information before we just jump into writing an equation.

▶ In $\triangle FMA$, $AF = 150$ cm, $FM = 165$ cm, and m$\angle A = 70°$. How many triangles can be constructed?

▶ The arc drawn from point F is longer than AF so 1 triangle can be constructed.

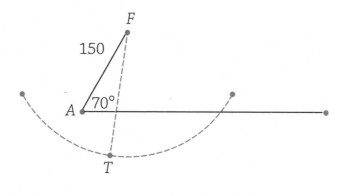

This is the last example involving an acute angle:

▶ In $\triangle FMA$, $AF = 150$ cm, $FM = 165$ cm, and m$\angle A = 70°$. How many triangles can be constructed?

▶ The shortest segment from F that can form a triangle has length $150 \sin(70) \approx 140.95$ cm. Because the length of FM is between $150 \sin(70)$ and 150 (the length of AF), there are two triangles that can be constructed.

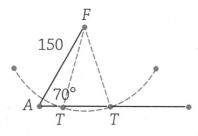

If the angle is obtuse, a triangle can be constructed only when the side opposite the given angle is longer than the given side.

EXAMPLE

▶ In $\triangle MPG$, $MP = 280$ m, $MG = 275$ cm, and $m\angle P = 115°$. How many triangles can be constructed?

▶ With $\angle P$ being an obtuse angle, MG must be longer than MP to create a triangle. Because MG < MP, there are 0 triangles that can be constructed.

Law of Cosines

The Law of Sines is used when the information available for the triangle fits one of the patterns ASA, AAS, or SSA. When the information available for the triangle fits the pattern SSS or SAS, the **Law of Cosines** is used to find the missing information about the triangle. The formula for the Law of Cosines is

$$p^2 = q^2 + r^2 - 2\,qr\,cos(P)$$

The key to this formula is to realize that when the information fits the **SAS** pattern, it is the third side of the triangle that is first determined. When the information is of the **SSS** pattern, it is an angle that is first determined. In the formula for the Law of Cosines, the side and the angle opposite that side are at the beginning and end of the formula.

EXAMPLE

▶ In $\triangle DUE$, $DE = 100$ cm, $UE = 80$ cm, and $m\angle E = 51°$. Find, to the nearest tenth of a centimeter, the length of $\overline{DU}$.

▶ $\overline{DU}$ is also known as side e. The information given for $\triangle DUE$ fits the SAS pattern, so use the Law of Cosines to get

$$e^2 = 100^2 + 80^2 - 2(100)(80) \cos(51)$$

▶ Compute to get e^2 with your calculator (approximately 6330.87). Take the square root of the value on the calculator screen to determine $e = DU = 79.6$ cm.

This next example asks us to find the measure of an angle of a triangle given the lengths of the three sides.

EXAMPLE

▶ Given $\triangle SPY$ with $SP = 27.3$ m, $PY = 17.1$ m, and $YS = 21.9$ m, find $m\angle Y$ to the nearest tenth of a degree.

▶ The information provided for $\triangle SPY$ fits the SSS pattern. Use the Law of Cosines and note that $\overline{SP}$ is opposite $\angle Y$ in this triangle.

$$27.3^2 = 17.1^2 + 21.9^2 - 2(17.1)(21.9) \cos(Y)$$

$$745.29 = 292.41 + 479.61 - 748.98 \cos(Y)$$

$$745.29 = 772.02 - 748.98 \cos(Y)$$

▶ The temptation is to subtract 748.98 from 772.02, but this is not mathematically correct. In the same way that one cannot simplify $5 - 3x$ because the terms are not "like" terms, 540.25 and 503.04 $\cos(Y)$ are not like terms.

▶ Subtract 772.02 from both sides of the equation.

$$-26.73 = -748.98 \cos(Y)$$

▶ Solve for $\cos(Y)$.

$$\cos(Y) = \frac{-26.73}{-748.98} \text{ so } m\angle Y = \cos^{-1}\left(\frac{-26.73}{-748.98}\right) = 88.0°$$

The Law of Sines or the Law of Cosines can be used to determine m∠P and m∠S if you are directed to do so, although the Law of Sines would be less work computationally.

The Law of Sines and the Law of Cosines are used in the resolution of vectors. Vectors are represents as directed line segments (they have length and direction) and the sum of two vectors is accomplished using the parallelogram method. That is, the vectors are placed together at their "tails" (endpoints) with the direction of the vector indicated with an arrow, as shown in the accompanying diagram.

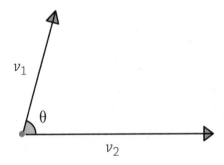

A **parallelogram** is constructed from these two sides, and the diagonal of the parallelogram drawn from the original set of tails is called the **resultant**, or **net impact**, of the two vectors.

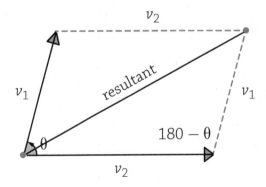

EXAMPLE

▶ Two forces with magnitudes 100 N and 120 N act on an object with an angle of 72° between the two forces. (A Newton is the international unit of measure for force. It is equal to one kilogram-meter per second squared.) Determine the magnitude of the resultant force, correct to the nearest tenth of a newton.

▶ Construct the vector diagram.

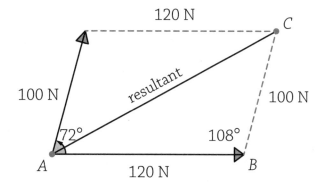

▶ △ABC displays SAS information. Use the Law of Cosines to determine the magnitude of the resultant, r.

$$r^2 = 100^2 + 120^2 - 2(100)(120)\cos(72)$$
$$r^2 = 16983.6$$
$$r = 130.3 \text{ N}$$

We are now given the magnitude of the resultant and asked to determine the measure of the angle between the original forces.

EXAMPLE

▶ Two forces of 150 lbs and 180 lbs act on an object with a resultant force of 160 lbs. Determine, to the nearest tenth of a degree, the angle between the two forces.

▶ Draw the vector diagram for this problem.

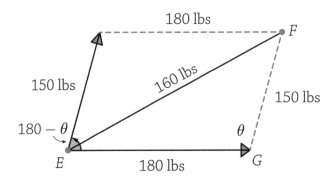

▶ $\triangle EFG$ displays SSS information. Use the Law of Cosines to determine the measure of angle θ.

$$160^2 = 150^2 + 180^2 - 2(150)(180)\cos(\theta)$$

$$25{,}600 = 22{,}500 + 32{,}400 - 54{,}000\cos(\theta)$$

$$25{,}600 = 54{,}900 - 54{,}000\cos(\theta)$$

$$-29{,}300 = -54{,}000\cos(\theta)$$

$$\cos(\theta) = \frac{-29{,}300}{-54{,}000}$$

$$\theta = \cos^{-1}\left(\frac{-29{,}300}{-54{,}000}\right) = 57.14°$$

▶ The angle between the two forces is the supplement of this angle, $180° - 57.14° = 122.9°$.

This last example involves both the Law of Cosines and the Law of Sines.

▶ Two forces of 60 N and 90 N act on an object with an angle of 110° between them. Find, to the nearest tenth of a degree, the measure of the angle formed between the resultant and the larger force.

▶ Draw the vector diagram.

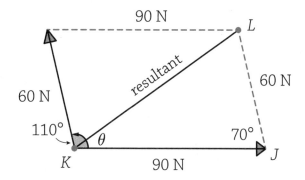

▶ $\triangle JKL$ displays SAS information. Use the Law of Cosines to determine the magnitude of the resultant, r.

$$r^2 = 60^2 + 90^2 - 2(60)(90)\cos(110)$$

$$r^2 = 3600 + 8100 - 10800\cos(110)$$

$$r^2 = 11700 - 10800\cos(110)$$

$$r^2 = 15393.8$$

$$r = 124.072$$

▶ Because this is an intermediate step in the problem, you will use the full number of decimal places from your calculator in the next step. In writing the equation, rather than copy the approximation for r into the equation, we will write r.

▶ The Law of Sines can be used to compute the value of θ.

$$\frac{r}{\sin(70)} = \frac{60}{\sin(\theta)}$$

$$r\sin(\theta) = 60\sin(70)$$

$$\sin(\theta) = \frac{60\sin(70)}{r}$$

$$\theta = \sin^{-1}\left(\frac{60\sin(70)}{r}\right) = 27.0°$$

EXERCISES

EXERCISE 9-1

Find the length of the indicated side. Round answers to the nearest tenth.

1.

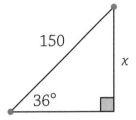

2.

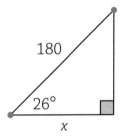

3.

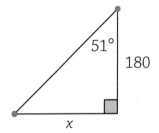

EXERCISE 9-2

Find, to the nearest tenth of a degree, the measure of θ.

1.

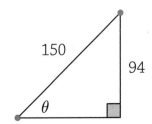

2.

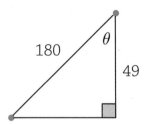

3.

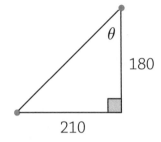

EXERCISE 9-3

Apply the Law of Sines to determine the missing measurements.

1. Meghan was hiking when she came to a flat region and saw a hill in the distance. Being a bit of a math lover, she happens to have an inclinometer in her backpack. She measures the angle of elevation to the top of the hill to be 21.3°. She walks directly toward the hill (in measured strides) another 600 feet. She measures the angle of elevation to the top of the hill to be 47.3°. She knows that she can now calculate the height of the hill when she gets back home. Determine, to the nearest foot, the height of the hill.

2. In $\triangle ABC$, $AB = 50$ in, $AC = 40$ in, and $m\angle A = 49°$. Find the area of $\triangle ABC$ to the nearest square inch.

3. Given $\triangle WHY$ with $WH = 120$ ft., $HY = 130$ ft., and $m\angle H = 141°$, find the area of $\triangle WHY$ to the nearest square foot.

4. The area of $\triangle KLM$ is 8693 square centimeters. If $KL = 140$ cm and $KM = 160$ cm, find the measure of angle K to the nearest degree.

5. The area of obtuse $\triangle GHJ$ is 71,365 square centimeters. If $GH = 520$ cm and $GJ = 410$ cm, find, to the nearest degree, the measure of the obtuse angle G.

6. Given $\triangle ACF$ with $AC = 30$ cm, $AF = 44$ cm, and m$\angle C = 63.2°$, find m$\angle F$ to the nearest tenth of a degree.

7. Given $\triangle KLM$ with $KL = 73.2$ ft, $LM = 101.4$ ft, and m$\angle K = 57°$, find m$\angle M$ to the nearest tenth of a degree.

8. Given $\triangle QRS$ with m$\angle Q = 35.6°$, m$\angle R = 43.6°$, and $QR = 17.1$ mm, find QS and RS correct to the nearest tenth of a centimeter.

9. Given $\triangle YES$ with m$\angle Y = 46.9°$, m$\angle E = 76.3°$, and $YS = 78.1$ mm, find ES and EY correct to the nearest tenth of a centimeter.

EXERCISE 9-4

Determine the number of triangles that can be constructed from the given information.

1. In $\triangle KAT$, $KA = 78$ in, $KT = 42$ in, and m$\angle A = 40°$. How many triangles, if any, can be constructed?

2. In $\triangle DAY$, $DA = 128$ ft., $YD = 125$ ft., and m$\angle A = 70°$. How many triangles, if any, can be constructed?

3. In $\triangle DAT$, $TA = 28$ ft., $TD = 44$ ft., and m$\angle A = 40°$. How many triangles, if any, can be constructed?

4. In $\triangle TEN$, $EN = 38$ cm., $TE = 44$ cm., and m$\angle E = 140°$. How many triangles, if any, can be constructed?

EXERCISE 9-5

Apply the Law of Cosines to determine the missing measurements.

1. Given $\triangle ENC$, $CN = 30$ cm., $NE = 45$ cm., and m∠$N = 73°$, find EC to the nearest centimeter.

2. Given $\triangle XYZ$, $XY = 76$ cm., $YZ = 54$ cm., and $YX = 58$ cm, find m∠X to the nearest tenth of a degree.

3. Given $\triangle PTS$, $PT = 37.6$ cm., $PS = 45.4$ cm., and $TS = 13.9$ cm, find m∠T to the nearest tenth of a degree.

EXERCISE 9-6

Apply the Law of Sines or Law of Cosines to determine the missing measurements.

1. Vectors with magnitudes 73 N and 65 N act on an object at an angle of 40° to each other. Find, to the nearest tenth of a newton, the magnitude of the resultant force.

2. Vectors with magnitudes 120 N and 150 N act on an object with a resultant force of 175 N. Find, to the nearest degree, the angle between the two original forces.

3. Vectors with magnitudes 250 N and 270 N act on an object with a resultant force of 295 N. Find, to the nearest tenth of a degree, the angle between the resultant and the larger force.

Flashcard App

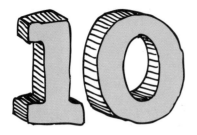

10 Trigonometric Functions

MUST ⚡ KNOW

⚡ The unit circle enables us to apply trigonometric functions to angles involving the radii of a circle.

⚡ Radian measures for angles have no units.

⚡ Trigonometric identities involve at least one trigonometric function that is true for all values of the variable(s) for which both sides of the equation are defined.

⚡ Proving trig identities allows us to simplify expressions to where they only contain sine and cosine ratios.

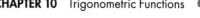

e now get the opportunity to look at the trigonometric functions as they are used outside of a geometric perspective. To do so, we apply the basic right triangle inside a circle to get the basic definitions and relationships established, then get to examine a number of applications.

Unit Circle

The unit circle is the circle centered at the origin with a radius of 1. We'll begin our examination in the first quadrant to identify some basic relationships and then will expand the study to the entire plane.

The First Quadrant

When the right triangle is placed inside this circle, as shown in the next figure, important relationships are determined.

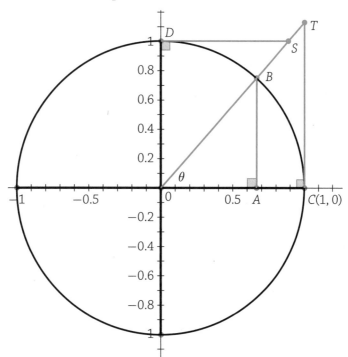

With the acute angle at the origin measuring θ, we use the trigonometric relationships from right triangles to determine that $OA = \cos(\theta)$ and $AB = \sin(\theta)$. The coordinates of point B are $(\cos(\theta), \sin(\theta))$. With the equation of the unit circle as $x^2 + y^2 = 1$, the coordinates for point B translate to the most important trigonometric identity $\cos^2(\theta) + \sin^2(\theta) = 1$. (Please note that $\cos^2(\theta)$ really means the square of $\cos(\theta)$.) This is the first of three Pythagorean identities.

$\triangle OAB \sim \triangle OCT$ because they share angle θ and they each have a right angle. Recall that corresponding sides of similar triangles are in proportion. We use this to determine three very important relationships.

First, there is the proportion $\dfrac{AB}{OA} = \dfrac{CT}{OC}$. We use the fact that $AB = \sin(\theta)$, $OA = \cos(\theta)$, and $OC = 1$. The proportion becomes $\dfrac{\sin(\theta)}{\cos(\theta)} = \dfrac{CT}{1} = CT$. However, $\overline{CT}$ is a segment on the line drawn tangent to the circle at point C. It is for this reason that $CT = \tan(\theta)$. This gives us a trigonometric definition for the tangent ratio, $\tan(\theta) = \dfrac{\sin(\theta)}{\cos(\theta)}$.

The second important proportion from these similar triangles is $\dfrac{OB}{OA} = \dfrac{OT}{OC}$. $OB = OC = 1$ and $OA = \cos(\theta)$. This yields the relationship $\dfrac{1}{\cos(\theta)} = OT$. $\overline{OT}$ is a segment on the line that passes through the circle twice (extend the ray through O to get the second point of intersection). Such a line is called the secant line. For this reason, OT, which is the reciprocal of $\cos(\theta)$, is called the secant of θ, abbreviated $\sec(\theta)$. We know that $\sec(\theta) = \dfrac{1}{\cos(\theta)}$.

$\triangle OCT$ is a right triangle. Applying the Pythagorean theorem, we get $OC^2 + CT^2 = OT^2$. Substituting the new trigonometric values just learned, this equation becomes the second Pythagorean identity, $1 + \tan^2(\theta) = \sec^2(\theta)$.

$\overline{SD} || \overline{OC}$ because they are both perpendicular to the y-axis. Therefore, the measure of $\sphericalangle DSO$ must also be θ (as it is an alternate interior angle to

∢BOA). We now have another pair of similar triangles, $\triangle OAB \sim \triangle SDO$. We can use the ratio of corresponding sides to yield two new trigonometric functions. First, there is $\dfrac{SD}{OD} = \dfrac{OA}{AB}$. With $OD = 1$ and $\overline{SD}$ being the line drawn tangent from the angle complementary to angle θ, SD is called the cotangent function, $\cot(\theta)$, so that $\cot(\theta) = \dfrac{\cos(\theta)}{\sin(\theta)} = \dfrac{1}{\tan(\theta)}$. In the same token, OS is called the cosecant of θ, $\csc(\theta)$, and $\csc(\theta) = \dfrac{1}{\sin(\theta)}$. Finally, $\triangle SDO$ is a right triangle and $OD^2 + DS^2 = OS^2$ or $1 + \cot^2(\theta) = \csc^2(\theta)$, the third Pythagorean identity.

Let's apply these new terms.

EXAMPLE

▶ Given acute angle A with $\sin(A) = \dfrac{3}{5}$, determine the values of the $\cot(A)$.

▶ Use the Pythagorean identity $\cos^2(A) + \sin^2(A) = 1$ to get $\cos^2(A) + \left(\dfrac{3}{5}\right)^2 = 1$. Solve this to get $\cos(A) = \dfrac{4}{5}$.

▶ Therefore, $\cot(A) = \dfrac{\cos(A)}{\sin(A)} = \dfrac{\frac{4}{5}}{\frac{3}{5}} = \dfrac{4}{3}$.

Here is another example.

EXAMPLE

▶ Given acute angle B with $\tan(B) = \dfrac{1}{2}$, determine the values of $\sin(B)$.

▶ Cotangent is the reciprocal of tangent, so $\cot(B) = 2$. Use the Pythagorean identity $1 + \cot^2(B) = \csc^2(B)$ to get $1 + 2^2 = \csc^2(B)$, which in turn tells us that $\csc(B) = \sqrt{5}$.

▶ Therefore, $\sin(B) = \dfrac{1}{\sqrt{5}} = \dfrac{\sqrt{5}}{5}$.

Beyond the First Quadrant

Angles of rotation become an issue when angles move beyond the first quadrant. When drawn in standard position, the initial side of the angle is the positive x-axes. If the terminal side is drawn in a counterclockwise manner from the initial side, the angle is said to have positive measure; if drawn in a clockwise manner, the angle has negative measure. Angles whose terminal sides are the same ray are called coterminal angles. For example, an angle with measure 130° and an angle with measure −230° are co-terminal. These same angles are coterminal with angles having measures 490°, 850°, 1210°, −590°, and −950°. It is possible to have angles with measures greater than 360° (think about a car spinning on ice—"doing a 360" = more than one revolution yields an angle that is more than 360°.)

Let's try this in an example.

▶ Find two angles, one with positive measure and one with negative measure, which are coterminal with an angle whose measure is 215°.

▶ $360° + 215° = 575°$ will be an angle that is coterminal with 215°. $215° − 360° = −145°$ will also be an angle that is coterminal with 215°.

When the terminal side of θ goes beyond the first quadrant, the rules for opposite, adjacent, and hypotenuse need to be reconsidered. For example, when $\theta = 90°$, the coordinates for point B are (0, 1). Therefore, $\cos(90) = 0$ and $\sin(90) = 1$. $\csc(90)$ is also 1, while $\sec(90)$ and $\tan(90)$ are both undefined. (Stop a moment to think about this from an algebraic perspective, $\dfrac{1}{0}$ is undefined. From a geometric perspective, $\overrightarrow{OD} \,||\, \overleftrightarrow{CT}$ so there are no points of intersection.) Angles that terminate on one of the axes are called quadrantal angles.

θ	sin(θ)	cos(θ)	tan(θ)	csc(θ)	sec(θ)	cot(θ)
0°	0	1	0	Undefined	1	Undefined
90°	1	0	Undefined	1	Undefined	0
180°	0	−1	0	Undefined	−1	Undefined
270°	−1	0	Undefined	−1	Undefined	0

If θ terminates within one of the quadrants, reflexive symmetry is used from a corresponding point in the first quadrant to determine the values of the trigonometric values.

In each case, when point B is reflected back into the first quadrant, a triangle congruent to $\triangle OAB$ is formed. The acute angle in the first quadrant is called the **reference angle** for θ. It is imperative that you notice that the reference angle is always the acute angle the terminal side of θ makes with the x-axis.

Here is an example from the second quadrant.

EXAMPLE

▶ Find the reference angle for 125°.

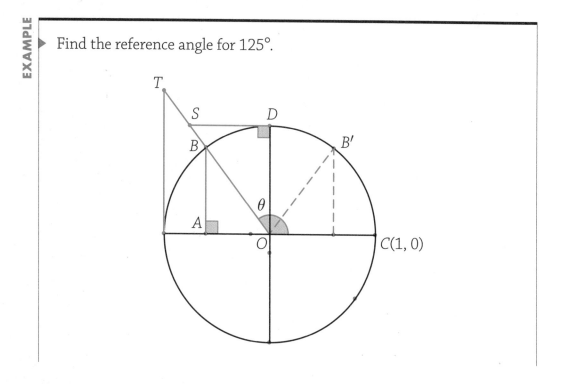

An angle with measure 125° terminates in the second quadrant. The reference angle is formed with the negative x-axis (a 180° angle), so the measure of the reference angle is $180° - 125° = 55°$.

Here is an example from the third quadrant.

Find the reference angle for 219°.

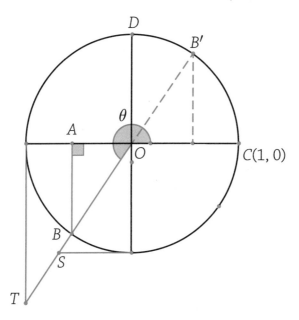

An angle with measure 219° terminates in the third quadrant. The reference angle is formed with the negative x-axis (180°), so the reference angle is $219° - 180° = 39°$.

Let's look at an example from the fourth quadrant.

Find the reference angle for 310°.

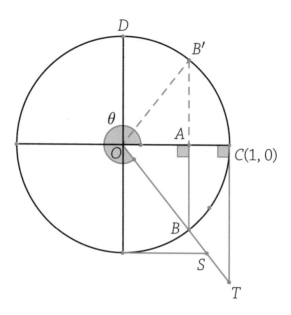

▶ An angle with measure 310° terminates in the fourth quadrant. The reference angle is formed with the positive x-axis (360°), so the reference angle is $360° - 310° = 50°$.

Now that we can work back to an acute angle, we are able to express the trigonometric value of a function of any angle in terms of that acute angle.

▶ Express $\sin(-219°)$ as a function of a positive acute angle.

▶ The positive acute angle is always the reference angle. In this case, the acute angle will have measure 39°. In the third quadrant, the y-coordinate of B is negative.

▶ Therefore, $\sin(219°) = -\sin(39°)$.

Let's take a look at another example from the fourth quadrant.

▶ Express $\cos(310°)$ as a function of a positive acute angle.

▶ Since cosine represents the x-coordinate of B, $\cos(310°)$ will be positive. With a reference angle of $50°$, $\cos(310°) = \cos(50°)$.

When working with the tangent and cotangent functions, we need to take into account the signs of both the first and second coordinates.

▶ Express $\tan(125°)$ as a function of a positive acute angle.

▶ Tangent is the ratio of the sine to the cosine (or the ratio of the y-coordinate to the x-coordinate). In the second quadrant, this ratio will be negative. Therefore, $\tan(125°) = -\tan(55°)$.

We'll do one more example, but this time with an angle that has a negative value.

▶ Express $\sin(-129°)$ as a function of a positive acute angle.

▶ The reference angle for $-129°$ is $51°$. The angle $-129°$ terminates in the third quadrant, so the y-coordinate will be negative. Therefore, $\sin(-129°) = -\sin(51°)$.

Radian Measure

By this time in your education, you probably have had many experiences dealing with different units of measurement. There is standard (inches, feet, yards, miles) versus metric (centimeter, meter, kilometer), as well as Fahrenheit versus Celsius.

The use of the degree as a measure of angle measure goes back to the Babylonians and a calendar with 360 days (a 5-day religious celebration at the end of the year kept their calendar relatively accurate for their short time as a power in history). A more accurate angle measure (despite its lack of usage among the general public) is the radian. By definition, the radian measure of a <central> angle is the ratio of the arc formed by the angle to the length of the radius. A key piece to this definition is that the radian has no units. Arc length and radius will be measured in the same units (whether standard or metric) and so will cancel each other within the ratio.

The conversion between degrees and radians is best considered when using a circle with radius 1 (although this is not necessary, as a dilation will create proportionally large radii and arcs without changing angles). A complete revolution of the circle is 360°, but it is also an arc of length 2π radians. Half a revolution is 180° or π radians. It is usually this reduced set of numbers that is used to convert from one measurement to the other:

$$\frac{\pi}{180} = \frac{\text{radian measure}}{\text{degree measure}}$$

Let's practice the conversion between the two units.

EXAMPLE

▶ Convert 60° to radians.

▶ $\dfrac{r}{60} = \dfrac{\pi}{180}$ becomes $r = \dfrac{60\pi}{180} = \dfrac{\pi}{3}$.

Of course, this means that twice 60° will be twice $\dfrac{\pi}{3}$ so that 120° is equivalent to $\dfrac{2\pi}{3}$, while half of 60° will be half $\dfrac{\pi}{3}$, so 30° is equivalent to $\dfrac{\pi}{6}$.

▶ Convert $\dfrac{7\pi}{15}$ radians to degrees.

▶ $\dfrac{\frac{7\pi}{15}}{d} = \dfrac{\pi}{180}$ becomes $\pi d = \left(\dfrac{7\pi}{15}\right)180$ so that $\pi d = 84\pi$ and $d = 84°.$

We should do one more problem to help solidify the relationship between degrees and radians.

▶ How many degrees correspond to 1 radian?

▶ The answer to this question is very simple, $\dfrac{180}{\pi}$. That seems very easy, but it does not provide a good feel for the relationship.

▶ Converting this fraction to a decimal tells us that 1 radian is approximately 57.3°.

If θ is the radian measure of the central angle, r is the length of the radius, and s is the length of the arc, the formula $\theta = \dfrac{s}{r}$ becomes $s = r\,\theta.$

▶ A tire with a radius of 14 inches makes 2500 revolutions. How far did the tire travel?

▶ A point on the tire will travel a linear distance of $14(2\pi) = 28\pi$ inches after one revolution. After 2500 revolutions, the point on the tire will have traveled $70{,}000\pi$ inches (18,326 feet or 3.47 miles).

The area of a sector of a circle is a proportional part of the area of the entire circle. Given a sector with a central angle with measure θ radians,

the area of the sector, A_s, can be computed with the proportion $\dfrac{\theta}{2\pi} = \dfrac{A_s}{\pi r^2}$.
Solving this for A_s, $A_s = \dfrac{1}{2}r^2\theta$.

EXAMPLE

▶ Determine the area of the sector of a circle formed if the radius of the circle is 12 cm and the measure of the central angle is 100°.

▶ We need to convert the angle from degrees to radians before we can apply the formula. 100° is equivalent to $\dfrac{5\pi}{9}$ radians. The area of the sector is $\dfrac{1}{2}(12)^2\dfrac{5\pi}{9} = 40\pi$ sq. cm.

Notice that the formula for the area of the sector of a circle can be written as $A_s = \dfrac{1}{2}r(r\theta) = \dfrac{1}{2}rs$, where s is the length of the arc that forms a boundary for the sector.

EXAMPLE

▶ In a circle with radius r cm, a central angle of θ radians forms a sector with perimeter $40 + 12\pi$ cm and an area of 120π sq. cm. Determine the length of the radius of the circle and the measure of the central angle.

▶ The perimeter is formed by both radii and the arc along the circle. That is, $2r + r\theta = 40 + 12\pi$. The area of the sector is given by the equation $\dfrac{1}{2}r^2\theta = 120\pi$ so that $r^2\theta = 240\pi$.

▶ We can solve this system of equations by solving for θ in the second equation, $\theta = \dfrac{240\pi}{r^2}$, and substituting this into the first equation, $2r + r\left(\dfrac{240\pi}{r^2}\right) = 40 + 12\pi$. Reduce the fraction to get

$2r + \dfrac{240\pi}{r} = 40 + 12\pi$. Multiply both sides of the equation by r to get

$$2r^2 + 240\pi = (40 + 12\pi)r$$

$$2r^2 - (40 + 12\pi)r + 240\pi = 0$$

▶ Now that is just out and out an ugly quadratic equation. Or is it? Divide by 2.

$$r^2 - (20 + 6\pi)r + 120\pi = 0$$

▶ We are looking for two numbers that add to be $20 + 6\pi$ and multiply to be 120π. The answer is right in front of us, 20 and 6π.

$$(r - 20)(r - 6\pi) = 0$$

$$r = 20, 6\pi$$

▶ If $r = 20$, then $\theta = \dfrac{240\pi}{20^2} = \dfrac{240\pi}{400} = \dfrac{3\pi}{5}$.

▶ If $r = 6\pi$, then $\theta = \dfrac{240\pi}{(6\pi)^2} = \dfrac{240\pi}{36\pi^2} = \dfrac{20}{3\pi}$.

▶ For those of you who looked at the perimeter of the sector and wrote down the first part of the solution, you now know to be more careful!

BTW

Something to learn from this problem is that, even though the vast majority of the examples and exercises that you will encounter will have angles (in radian measure) of the form $\dfrac{a\pi}{b}$, a radian value is a real number and **any** real number might be the measure of the angle. So the moral of the story here is to be careful in your steps when solving equations involving the angle in radian measure but be ready to accept what that answer could look like.

Trigonometric Identities

Through the work done to this point, we have established a number of trigonometric identities:

- **Quotient Identities**

$$\tan(\theta) = \frac{\sin(\theta)}{\cos(\theta)} \qquad \cot(\theta) = \frac{\cos(\theta)}{\sin(\theta)}$$

- **Reciprocal Identities**

$$\sec(\theta) = \frac{1}{\cos(\theta)} \qquad \csc(\theta) = \frac{1}{\sin(\theta)} \qquad \cot(\theta) = \frac{1}{\tan(\theta)}$$

- **Cofunction Identities**

$$\sin\left(\frac{\pi}{2} - \theta\right) = \cos(\theta) \qquad \sec\left(\frac{\pi}{2} - \theta\right) = \csc(\theta) \qquad \tan\left(\frac{\pi}{2} - \theta\right) = \cot(\theta)$$

- **Pythagorean Identities**

$$\cos^2(\theta) + \sin^2(\theta) = 1 \quad 1 + \tan^2(\theta) = \sec^2(\theta) \quad 1 + \cot^2(\theta) = \csc^2(\theta)$$

- **Reflection Identities** (all written with radian measure in mind)

$$\sin(\pi - \theta) = \sin(\theta) \qquad \cos(\pi - \theta) = -\cos(\theta)$$

$$\sin(\pi + \theta) = -\sin(\theta) \qquad \cos(\pi + \theta) = -\cos(\theta)$$

$$\sin(2\pi - \theta) = -\sin(\theta) \qquad \cos(2\pi - \theta) = \cos(\theta)$$

- **Rotational Identities**

$$\sin(-\theta) = -\sin(\theta) \qquad \cos(-\theta) = \cos(\theta) \qquad \tan(-\theta) = -\tan(\theta)$$

There are a number of other important formulas that you will need to be comfortable with applying. The derivation for $\sin(A + B)$ and $\cos(A + B)$ can be found online if you are interested in looking at them.

■ **Addition and Subtraction Formulas**

$$\sin(A + B) = \sin(A)\cos(B) + \sin(B)\cos(A)$$

$$\sin(A - B) = \sin(A)\cos(B) - \sin(B)\cos(A)$$

$$\cos(A + B) = \cos(A)\cos(B) - \sin(A)\sin(B)$$

$$\cos(A - B) = \cos(A)\cos(B) + \sin(A)\sin(B)$$

$$\tan(A + B) = \frac{\sin(A + B)}{\cos(A + B)} = \frac{\tan(A) + \tan(B)}{1 - \tan(A)\tan(B)}$$

$$\tan(A - B) = \frac{\sin(A - B)}{\cos(A - B)} = \frac{\tan(A) - \tan(B)}{1 + \tan(A)\tan(B)}$$

■ **Double Angle**

$$\sin(2A) = 2\sin(A)\cos(A)$$

$$\cos(2A) = \cos^2(A) - \sin^2(A)$$

$$= 2\cos^2(A) - 1$$

$$= 1 - 2\sin^2(A)$$

$$\tan(2A) = \frac{2\tan(A)}{1 - \tan^2(A)}$$

Let's work through a few examples of these identities.

▶ Given $\sin(A) = \dfrac{-60}{61}$ with $180° < A < 270°$ and $\cos(B) = \dfrac{12}{13}$, $270° < B < 360°$. Find $\sin(A + B)$.

▶ With A in the third quadrant, $\cos(A) < 0$. Use the Pythagorean identity $\cos^2(\theta) + \sin^2(\theta) = 1$ to get $\cos^2(A) + \left(\dfrac{-60}{61}\right)^2 = 1$ to show that $\cos(A) = \dfrac{-11}{61}$. Angle B is in the fourth quadrant.

▶ This tells us that $\sin(B) < 0$. Use the Pythagorean identity to determine $\sin(B) = \dfrac{-5}{13}$. We can compute the value of $\sin(A + B)$.

▶ $\sin(A + B) = \sin(A)\cos(B) + \sin(B)\cos(A) =$

$$\left(\frac{-60}{61}\right)\left(\frac{12}{13}\right) + \left(\frac{-11}{61}\right)\left(\frac{-5}{13}\right) = \frac{-665}{793}.$$

This problem involves a double angle:

EXAMPLE

▶ Given $\cos(Q) = \dfrac{-24}{25}$ with $\dfrac{\pi}{2} < Q < \pi$. Determine the value of $\tan(2Q)$.

▶ Angle Q terminates in the second quadrant. Use the Pythagorean identity to show $\sin(Q) = \dfrac{7}{25}$.

▶ Using the quotient identity, $\tan(Q) = \dfrac{-7}{24}$. We can now determine the value of $\tan(2Q)$.

$$\tan(2Q) = \frac{2\left(\dfrac{-7}{24}\right)}{1 - \left(\dfrac{-7}{24}\right)^2} = \frac{-336}{527}.$$

There is another set of identities we need to look at, the half-angle identities. The development for these identities is based on the two variations of the cosine of the double angle.

$$\cos(2A) = 2\cos^2(A) - 1$$

Let $x = 2A$. This implies that $A = \dfrac{x}{2}$. Substitute this into the equation.

$$\cos(x) = 2\cos^2\left(\frac{x}{2}\right) - 1$$

Solve for $\cos\left(\dfrac{x}{2}\right)$:

$$\cos^2\left(\frac{x}{2}\right) = \frac{\cos(x)+1}{2}$$

$$\cos\left(\frac{x}{2}\right) = \pm\sqrt{\frac{\cos(x)+1}{2}}$$

The sign in front of the radical is determined by the quadrant in which the half-angle terminates.

It can be shown in a similar fashion that $\sin\left(\dfrac{x}{2}\right) = \pm\sqrt{\dfrac{1-\cos(x)}{2}}$.

Since the tangent is the ratio of the sine value to the cosine value, we get

$$\tan\left(\frac{x}{2}\right) = \pm\sqrt{\frac{1-\cos(x)}{\cos(x)+1}}.$$

EXAMPLE

▶ Given $\sin(W) = \dfrac{-15}{17}$, $\pi < W < \dfrac{3\pi}{2}$, determine the value of $\cos\left(\dfrac{W}{2}\right)$.

▶ Use the Pythagorean identity to get $\cos(W) = \dfrac{-8}{17}$. With $\pi < W < \dfrac{3\pi}{2}$, we have $\dfrac{\pi}{2} < \dfrac{W}{2} < \dfrac{3\pi}{4}$ telling us the half-angle is in quadrant I and the cosine will be positive.

▶ Therefore, $\cos\left(\dfrac{W}{2}\right) = \sqrt{\dfrac{\dfrac{-8}{17}+1}{2}} = \sqrt{\dfrac{\dfrac{9}{17}}{2}} = \sqrt{\dfrac{9}{34}} = \dfrac{3}{\sqrt{34}}$.

Proving Trigonometric Identities

The process for proving trigonometric identities is similar to that of mathematical induction, in that we cannot apply any of the properties of equality to the problems because we do not know if the two sides of the statement are equal to each other. We can manipulate each side of the statement until we get to the point that the last line of each side of the statement is identical. Then we can claim the identity is true.

EXAMPLE

▶ Prove: $\csc(\theta) - \sin(\theta) = \cos(\theta)\cot(\theta)$

▶ Changing all statements to be in terms of $\sin(\theta)$ and $\cos(\theta)$ is often a safe way to proceed.

$$\frac{1}{\sin(\theta)} - \sin(\theta) = \cos(\theta)\left(\frac{\cos(\theta)}{\sin(\theta)}\right)$$

▶ Combine fractions:

$$\frac{1 - \sin^2(\theta)}{\sin(\theta)} = \frac{\cos^2(\theta)}{\sin(\theta)}$$

▶ Apply a variation of the Pythagorean identity: $\cos^2(\theta) = 1 - \sin^2(\theta)$

$$\frac{\cos^2(\theta)}{\sin(\theta)} = \frac{\cos^2(\theta)}{\sin(\theta)}$$

▶ Both sides of the statement are identical so the proof is complete.

Let's try another problem.

EXAMPLE

▶ Prove: $\dfrac{1 - \sin(\theta)}{\sec(\theta)} = \dfrac{\cos^3(\theta)}{1 + \sin(\theta)}$

▶ Change the problem to be in terms of $\sin(\theta)$ and $\cos(\theta)$.

$$\frac{1 - \sin(\theta)}{\dfrac{1}{\cos(\theta)}} = \frac{\cos^3(\theta)}{1 + \sin(\theta)}$$

▶ Simplify the complex fraction:

$$\cos(\theta)(1 - \sin(\theta)) = \frac{\cos^3(\theta)}{1 + \sin(\theta)}$$

▶ The next step might seem tricky, but it really is not. Notice that we no longer have a fraction on the left side of this statement. Also notice that the denominator on the right-hand side of the statement is one of the factors to $1 - \sin^2(\theta)$ *and* that the other factor is on the left side of the statement.

▶ Multiply the left side of the statement by $\dfrac{1 + \sin(\theta)}{1 + \sin(\theta)}$.

$$\cos(\theta)(1 - \sin(\theta))\frac{1 + \sin(\theta)}{1 + \sin(\theta)} = \frac{\cos^3(\theta)}{1 + \sin(\theta)}$$

▶ Combine the terms on the left side:

$$\frac{\cos(\theta)(1 - \sin^2(\theta))}{1 + \sin(\theta)} =$$

▶ Apply the Pythagorean identity.

$$\frac{\cos(\theta)\cos^2(\theta)}{1 + \sin(\theta)} =$$

▶ Multiply.

$$\frac{\cos^3(\theta)}{1 + \sin(\theta)} = \frac{\cos^3(\theta)}{1 + \sin(\theta)}$$

▶ The last lines on each side of the statement are identical. The identity is proved. You noticed that we didn't bother writing the right side of the statement until the last line to confirm the expressions were identical. It saves time, but if you want to continue to write that one down, feel free to do so.

The initial step, after changing all terms to $\sin(\theta)$ and $\cos(\theta)$, should be pretty clear.

EXAMPLE

▶ Prove: $\dfrac{1 - \cos(\theta)}{\sin(\theta)} + \dfrac{\sin(\theta)}{1 - \cos(\theta)} = 2\csc(\theta)$

▶ Change the right side of the expression in terms of $\sin(\theta)$.

$$\frac{1 - \cos(\theta)}{\sin(\theta)} + \frac{\sin(\theta)}{1 - \cos(\theta)} = \frac{2}{\sin(\theta)}$$

▶ Get a common denominator and add the fractions.

$$\frac{(1 - \cos(\theta))^2 + \sin^2(\theta)}{\sin(\theta)(1 - \cos(\theta))} = \frac{2}{\sin(\theta)}$$

▶ Expand the square.

$$\frac{1 - 2\cos(\theta) + \cos^2(\theta) + \sin^2(\theta)}{\sin(\theta)(1 - \cos(\theta))} =$$

▶ Apply the Pythagorean identity.

$$\frac{1 - 2\cos(\theta) + 1}{\sin(\theta)(1 - \cos(\theta))} =$$

▶ Remove the common factor of 2 in the numerator.

$$\frac{2(1 - \cos(\theta))}{\sin(\theta)(1 - \cos(\theta))} =$$

▶ Reduce the fraction.

$$\frac{2}{\sin(\theta)} = \frac{2}{\sin(\theta)}$$

▶ The identity is proven.

This last example looks much more complicated than it really is.

EXAMPLE

▶ Prove: $\cos^4(\theta) - \sin^4(\theta) = \cos(2\theta)$

▶ All terms are already in terms of $\sin(\theta)$ and $\cos(\theta)$. The left side of the expression is the difference of squares.

▶ Factor:

$$\left(\cos^2(\theta) - \sin^2(\theta)\right)\left(\cos^2(\theta) + \sin^2(\theta)\right) = \cos(2\theta)$$

▶ We have the Pythagorean identity, so one of the factors is equal to 1 and the other factor is one of the forms for the double-angle formula for the cosine function.

$$\cos(2\theta)(1) = \cos(2\theta)$$

▶ The identity is proven.

EXERCISES

EXERCISE 10-1

Convert each of these angle measures to radian mode.

1. $144°$

2. $135°$

3. $310°$

EXERCISE 10-2

Convert each of these angle measures to degree mode.

1. $\dfrac{5\pi}{8}$

2. $\dfrac{11\pi}{18}$

3. $\dfrac{7\pi}{12}$

EXERCISE 10-3

Given the acute angle Q with $\cos(Q) = \dfrac{15}{17}$, answer the following questions.

1. $\sec(Q)$

2. $\sin(Q)$

3. $\cot(Q)$

Given the acute angle W with $\sec(W) = \dfrac{7}{5}$, answer questions 4 to 6.

4. $\cos(W)$

5. $\sin(W)$

6. $\tan(W)$

EXERCISE 10-4

These questions relate to sectors of a circle.

1. In a circle with radius 20 cm, a central angle forms an arc with a length of 50 cm. Find the radian measure of the central angle.

2. A central angle with measure $\dfrac{6\pi}{5}$ forms an arc with a length of 24 cm. Find the length of the radius of the circle.

3. Determine the area of the sector formed when a central angle with a measure of $\dfrac{2\pi}{9}$ radians is drawn in a circle with a radius of 36 cm.

4. Determine the length of the radius of the circle and the measure of the central angle when a sector is formed with perimeter 240 inches and an area of 3200 sq. in.

EXERCISE 10-5

Find the measures of the reference angles for these questions.

1. 105°

2. 195°

3. 265°

4. 517°

EXERCISE 10-6

Express each trigonometric value as a function of a positive acute angle.

1. tan(207°)

2. sin(314°)

3. cos(138°)

EXERCISE 10-7

Find the specified functional value.

1. Given $\sin(\theta) = \dfrac{-15}{17}$ with θ terminating in quadrant III, find $\cos(\theta)$.

2. Given $\cos(\beta) = \dfrac{-24}{25}$ with β terminating in quadrant II, find $\tan(\beta)$.

3. Given $\tan(\alpha) = \dfrac{-3}{5}$ with β terminating in quadrant II, find $\csc(\alpha)$.

EXERCISE 10-8

Given $\sin(\alpha) = \dfrac{5}{13}$, $\dfrac{\pi}{2} < \alpha < \pi$, and $\cos(\beta) = \dfrac{-4}{5}$ $\pi < \beta < \dfrac{3\pi}{2}$, answer the following questions.

1. Find $\sin(\alpha + \beta)$.

2. Find $\cos(\alpha + \beta)$

3. Find $\tan(\alpha + \beta)$

4. Find $\sec(\alpha - \beta)$.

5. Find $\cos(2\beta)$.

6. Find $\cos\left(\dfrac{\beta}{2}\right)$.

7. Find $\sin\left(\alpha + \dfrac{\pi}{3}\right)$.

8. Find $\sin(2(\alpha + \beta))$.

9. Find $\tan\left(\dfrac{\alpha}{2}\right)$.

EXERCISE 10-9

Prove each trigonometric identity.

1. $\left(1 - \sin^2(\theta)\right)\left(1 + \tan^2(\theta)\right) = 1$

2. $\tan(\theta) + \dfrac{\cos(\theta)}{1 + \sin(\theta)} = \sec(\theta)$

3. $\dfrac{\sec(\beta) - 1}{\sec(\beta) + 1} = \dfrac{1 - \cos(\beta)}{1 + \cos(\beta)}$

4. $\dfrac{\tan(\alpha) + \tan(\beta)}{\cot(\alpha) + \cot(\beta)} = \tan(\alpha)\tan(\beta)$

5. $\sin(4x) = \cos(x)\left(4\sin(x) - 8\sin^3(x)\right)$

Flashcard App

Graphs and Applications of Trigonometric Functions

MUST ⚡ KNOW

⚡ Periodic phenomena are modeled with trigonometric functions.

⚡ The graphs of trig functions provide a visual representation of their properties.

⚡ Restricting the domain of a trigonometric function, we can create a 1:1 trig function that includes an inverse relationship.

Periodic phenomena are things you deal with every day. What time is it now? What will the time be 24 hours from now? One week from now? A month from now? A year from now? Phases of the moon (and therefore the tides) and the location of the stars in the night sky are probably what first got the early civilizations to study periodic phenomena, as their livelihoods and lives depended on this knowledge. EKGs show the beating of the heart, and in a healthy heart, they will show periodic tendencies. Alternating current is a periodic phenomenon. There are many other applications to periodic phenomena, some of which we will examine in this chapter.

Graphs of Trigonometric Functions

Imagine a circle centered at the origin. A particle is sitting on the circle at the point $(a, 0)$. The particle then begins to rotate in a counterclockwise direction while remaining on the circle. Furthermore, let's assume that it takes p time units (seconds, minutes, days, nanoseconds) for the particle to get back to its starting point.

If we track the y-coordinate of the particle as it rotates around the circle, plotting points with coordinates (time since motion began, y-coordinate), we would see the basic graph of the sine function (the sinusoid) graphed. The graph will begin to replicate itself after p time units; hence, we call p the period of the function. If we double the speed of the particle, the period will be cut in half. Similarly, if the speed of travel is one-third the original speed, the period will be three times the original value.

How far about the horizontal axis will the point reach? The radius of the circle is a, so when the particle is one-fourth of the way around the circle, the plotted point will be as far from the horizontal axis as it will get. When the particle is halfway around the circle, the plotted point will be back on the horizontal axis. When the particle is three-fourths of the way around the circle, the particle will once again be as far from the axis as it can be, but it will be on the reverse side of the axis as it was at the quarter point.

The graphed point returns to the horizontal axis when the particle returns to its starting point.

The scheme described in this last paragraph gives a good guide on how to get a reasonable sketch of any sine function of the form $f(x) = A\sin(Bx)$. The value A tells us the radius of the circle, and the value B helps determine the speed of the particle as it travels around the circle. If B is equal to 1, the time needed to travel the circle is 2π time units.

EXAMPLE

▶ Sketch the graph of $f(x) = 4\sin(2x)$.

▶ The circle in question has a radius of 4 units, and the period of the motion is $\dfrac{2\pi}{2} = \pi$ time units. The path of the particle is tracked with the coordinates (time, y-coordinate). The track begins at the point (0, 0). It works its way to $\left(\dfrac{\pi}{4}, 4\right)$ and then to $\left(\dfrac{\pi}{2}, 0\right)$. The point $\left(\dfrac{3\pi}{4}, -4\right)$ is the next major point encountered before the track returns to the horizontal axis at $(\pi, 0)$.

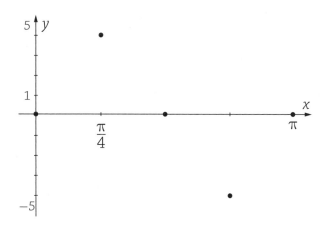

▶ Draw a smooth curve through these points to get the sketch.

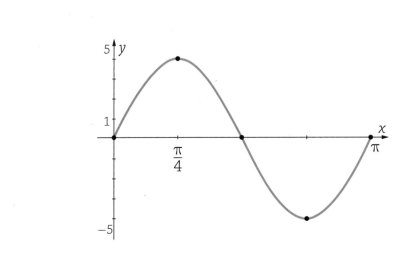

We can apply the same scenario of the particle traveling around the circle but track the x-coordinate rather than the y-coordinate.

▶ Sketch the graph of $g(x)=5\cos\left(\dfrac{1}{2}x\right)$.

▶ The circle in question has a radius of 5 units, and the period of the motion is $\dfrac{2\pi}{\dfrac{1}{2}}=4\pi$ time units. The path of the particle is tracked with the coordinates (time, y-coordinate). The track begins at the point (0, 5). It works its way to (π, 0) and then to (2π, −5). The point (3π, 0) is the next major point encountered before the track returns to the horizontal axis at (4π, 5).

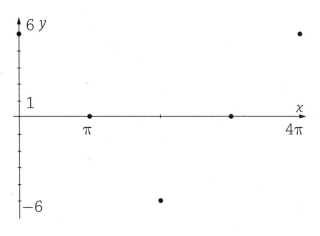

▶ Draw a smooth curve through these points.

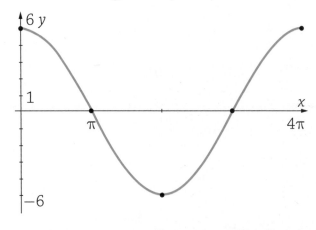

The two completed graphs show a competed cycle of the graph of a sine function and a cosine function. From here on, we'll use these as our basis for accuracy.

EXAMPLE

▶ Sketch the graph of $y = 4\sin(-2x)$.

▶ Let's follow the pattern that we established by going around the circle at the quarter points, knowing that we have to negate the input values as we do so.

▶ The circle in question has a radius of 4 units, and the period of the motion is $\dfrac{2\pi}{2} = \pi$ time units. The path of the particle is tracked with the coordinates (time, y-coordinate). The track begins at the point $(0, 0)$, It works its way to $\left(\dfrac{\pi}{4}, 4\right)$ and then to $\left(\dfrac{\pi}{2}, 0\right)$. The point $\left(\dfrac{3\pi}{4}, -4\right)$ is the next major point encountered before the track returns to the horizontal axis at $(\pi, 0)$. Negating the input values, the points we get are

$(0, 0)$, $\left(\dfrac{-\pi}{4}, 4\right)$, $\left(\dfrac{-\pi}{2}, 0\right)$, $\left(\dfrac{-3\pi}{4}, -4\right)$, and $(-\pi, 0)$.

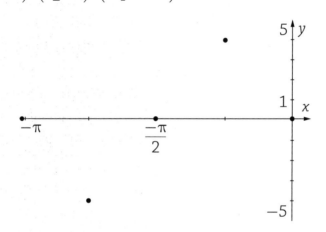

▶ Draw a smooth curve through these points.

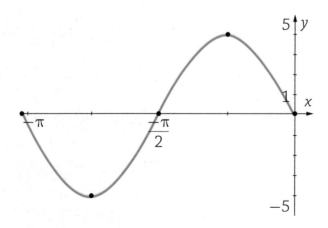

▶ It would be beneficial to compare the graph of this function to the graph of $y = 4\sin(2x)$. If we extend the viewing window to $x = \pi$, we see the following graph.

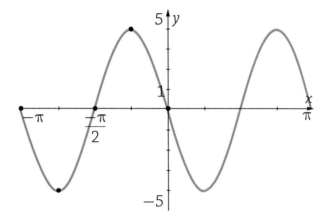

▶ As you can see, the graph of $y = 4\sin(-2x)$ is simply the reflection of the graph of $y = 4\sin(2x)$ over the x-axis. Do you see that in this case the graph of $y = 4\sin(-2x)$ $y = 4\sin(-2x)$ is simply the reflection of the graph of $y = 4\sin(2x)$ over the y-axis? This is because we know the identity $\sin(-\theta) = -\sin(\theta)$.

What happens with the graph of the cosine function when the argument of the function is negated?

EXAMPLE

▶ Sketch the graph of $g(x) = 5\cos\left(\dfrac{-1}{2}x\right)$.

▶ We know the reflexive property of the cosine function, $\cos(-\theta) = \cos(\theta)$, so the two graphs are exactly the same.

All we need to do is examine all the possible transformations that can occur. Hah! We need to examine what happens when we slide the graph up and down in the plane, reflect the graph over an axis, and stretch/compress the graph with respect to an axis. I believe you will find this fairly repetitive to what we just did. We need to determine a couple of pieces of information, and we will be good to go. We'll need to know where to start the graph, how far above the central axis it travels, how long the period is, and if it was flipped.

The basic forms for the equations of the graphs are $f(x) = A \sin(B(x - C)) + D$ and $g(x) = A \cos(B(x - C)) + D$. (Do you see that, with the exception of B, this is no different than examining the motions to the function $y = a\, p(x - h) + k$, where $p(x)$ is a basic function whose graph we know?)

Let's review:

- $|A|$ is the amplitude. It is the height of the graph from the central axis. Amplitude can be computed with the formula $\dfrac{\text{Max} - \text{Min}}{2}$.

- $|B|$ determines the period of the function with the formula $\dfrac{2\pi}{|B|}$.

- C is the horizontal translation; in trigonometry we call this the *phase shift*.

- D is the vertical translation.

Let's look at a few examples so that we can get a good feel for the process.

EXAMPLE

▶ Sketch the graph of $y = 3\sin\left(2\left(x - \dfrac{\pi}{3}\right)\right) + 1$.

▶ The circle in question has a radius of 3 units, and the period of the motion is $\dfrac{2\pi}{2} = \pi$ time units. The graph is translated to the right $\dfrac{\pi}{3}$

and up 1 unit. The path of the particle is tracked with the coordinates (time, y-coordinate).

▶ The track would normally begin at the point $(0, 0)$ and work its way to $\left(\frac{\pi}{4}, 4\right)$, $\left(\frac{\pi}{2}, 0\right)$, $\left(\frac{3\pi}{4}, -4\right)$, and $(\pi, 0)$. When we apply the translations to these values, we get the set of points $\left(\frac{\pi}{3}, 1\right)$, $\left(\frac{7\pi}{12}, 5\right)$, $\left(\frac{5\pi}{6}, 1\right)$, $\left(\frac{11\pi}{12}, -3\right)$, and $\left(\frac{4\pi}{3}, 1\right)$.

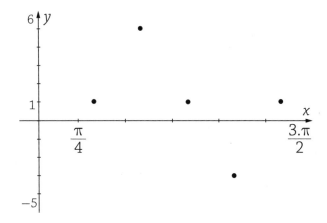

▶ Connect these points with a sinusoid.

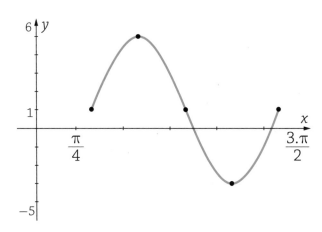

This next example uses a cosine function, but there is also a significant change in the equation.

▶ Sketch the graph of $y = 3\cos\left(2x - \dfrac{\pi}{3}\right) - 2$.

▶ We can determine that the amplitude of the function is 3 and the period will be π. The vertical translation is 2 down. The significant difference between this function and that of the last example is determining the horizontal translation.

▶ The format for the trigonometric function shows that the argument for the phase shift should be written as $(B(x - C))$, not $Bx{-}C$. We need

to rewrite $2x - \dfrac{\pi}{3}$ as $2\left(x - \dfrac{\pi}{6}\right)$ to determine the phase shift is $\dfrac{\pi}{6}$ to

the right. The key points for us to plot are $\left(\dfrac{\pi}{6}, 1\right)$, $\left(\dfrac{5\pi}{12}, -2\right)$, $\left(\dfrac{2\pi}{3}, -5\right)$,

$\left(\dfrac{11\pi}{12}, -2\right)$, and $\left(\dfrac{7\pi}{6}, 1\right)$.

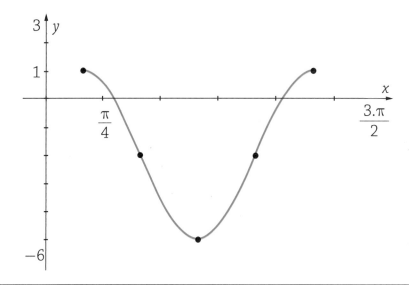

The next two examples often throw people for a loop because they are "different."

▶ Sketch the graph of $y = -2\sin(\pi(x - 1)) - 1$.

▶ The amplitude of the graph is 2; the graph is reflected over the x-axis and stretched from the axis by a factor of 2. The graph is translated down 1 unit. The graph is translated to the right 1 unit. The period of the graph is 2. The period is not 2π, just 2. Yes, we can graph trigonometric functions with a scale on the x-axis using integers.

▶ Now that we have that out of the way, let's find the coordinates of the key points that need to be plotted. The first value of x will be 1. The y-coordinate is –1. The remaining points are $\left(\dfrac{3}{2}, -3\right)$, $(-2, -1)$, $\left(\dfrac{5}{2}, 1\right)$, and $(3, 1)$. Plot the points and graph the sinusoid.

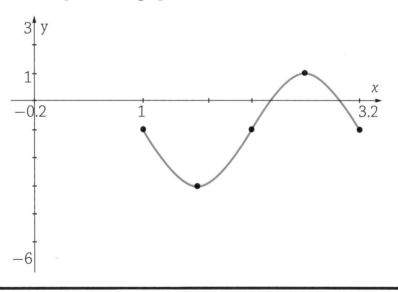

This next example goes so far as to have the input values not be multiples of π and the output values are multiples of π.

EXAMPLE

▶ Sketch the graph of $y = \pi \sin\left(\dfrac{2\pi}{3}(x+1)\right) + 2\pi$.

▶ The amplitude of the graph is π, the period is 3 $\left(\dfrac{2\pi}{\dfrac{2\pi}{3}} = 3\right)$, the phase

shift is 1 to the left, and the vertical translation is up 2π.

▶ The key points to plot are $(-1, 2\pi)$, $\left(\dfrac{-1}{4}, 3\pi\right)$, $\left(\dfrac{1}{2}, 2\pi\right)$, $\left(\dfrac{5}{4}, \pi\right)$, and

$(2, 2\pi)$. Plot the points and graph the sinusoid.

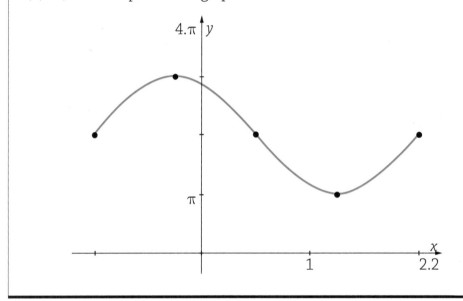

We need to take a look at the other functions before we move on. The tangent function is interesting in that there is no measure for amplitude, as its range is $(-\infty, \infty)$. The function is undefined whenever the input value is an odd multiple of $\dfrac{\pi}{2}$. What points can we work with to help with the sketch? If the input value is 0 or a multiple of π, the function takes on a value of 0. If the input value is an odd multiple of $\dfrac{\pi}{4}$, the output is either

1 or –1. When the function is not defined, we have asymptotes to help us sketch the graph.

▶ Sketch the graph of $y = \tan(x)$.

▶ Use the points $(0, 0)$, $\left(\dfrac{-\pi}{4}, -1\right)$, and $\left(\dfrac{\pi}{4}, 1\right)$ to sketch the graph.

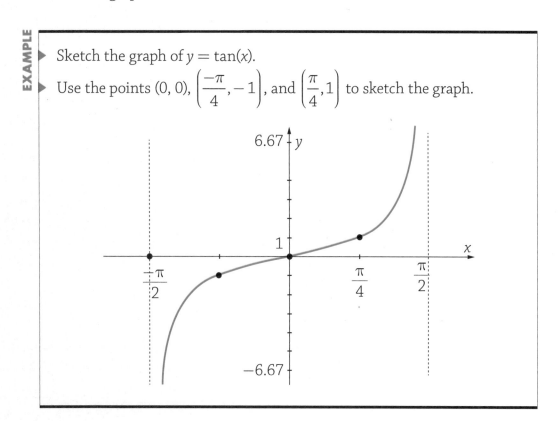

The cotangent function has a graph that is similar to that of the tangent function in that it is asymptotic with a period of π. The asymptotes occur at multiples of π.

▶ Sketch the graph of $y = \cot(x)$.

▶ Use the points $\left(\dfrac{\pi}{4}, 1\right)$, $(\pi, 0)$, and $\left(\dfrac{3\pi}{4}, -1\right)$ to sketch the graph.

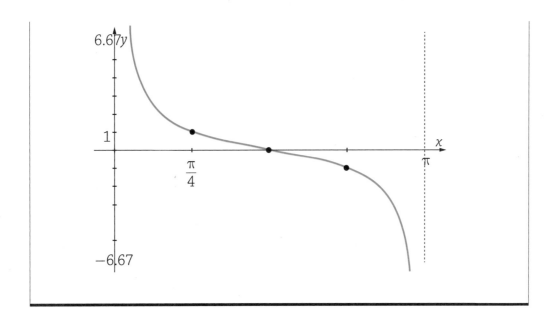

The graphs for the secant and cosecant functions are easily done if you first sketch the graph of the reciprocal function for reference.

▶ Sketch the graph of $y = 2\sec(x) + 1$.

▶ First sketch the graph of $y = 2\cos(x) + 1$. It is sketched as a dotted curve because it is a guideline for the real problem.

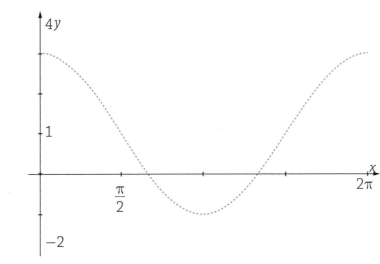

▶ The secant function is the reciprocal of the cosine function. The **cosine** function is equal to zero when $x = \dfrac{\pi}{2}, \dfrac{3\pi}{2}$, so the secant function will have asymptotes at these values of x. (Note that we did not look for where $y = 2\cos(x) + 1$ was equal to zero.) Now sketch the graph of $y = 2\sec(x) + 1$.

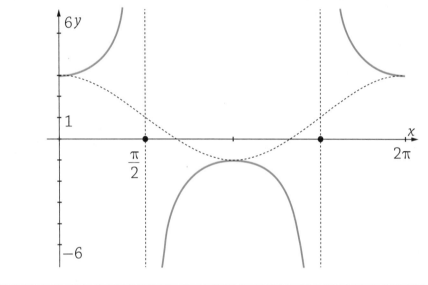

The same technique works with sketching graphs of the cosecant function.

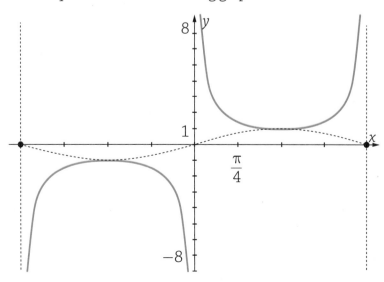

Inverse Trigonometric Functions

When you examine the graphs of $y = \sin(x)$, $y = \cos(x)$, and $y = \tan(x)$ over the interval $-2\pi \leq x < 2\pi$, you see that while each passes the vertical line test (indicating a function), they each fail the horizontal line test (the inverse is not a function). As you have seen done with the quadratic $y = x^2$, restricting the domain to allow for a $1 - 1$ function while covering the entire range of the function will allow for the creation of the inverse function. Convention has the restricted domain containing the value 0. The graphs of $y = \sin(x)$, $y = \cos(x)$, and $y = \tan(x)$ over an appropriate domain are shown.

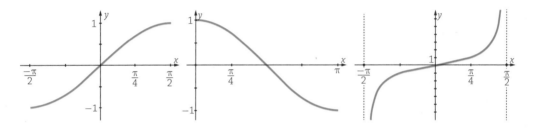

The restricted domain for $y = \sin(x)$ is $\dfrac{-\pi}{2} \leq x \leq \dfrac{\pi}{2}$, for $y = \cos(x)$ is $0 \leq x \leq \pi$, and for $y = \tan(x)$ is $\dfrac{-\pi}{2} < x < \dfrac{\pi}{2}$.

It is critical that you learn these domains because they impact the definition of the inverse trigonometric function and are the rules by which you will use your calculator for challenging problems.

Function	Domain	Range (Degrees)	Range (Radians)
$f^{-1}(x) = \sin^{-1}(x)$ or $\arcsin(x)$)	$-1 \leq x \leq 1$	$-90 \leq y \leq 90$	$\dfrac{-\pi}{2} \leq x \leq \dfrac{\pi}{2}$
$g^{-1}(x) = \cos^{-1}(x)$ or $\arccos(x)$	$-1 \leq x \leq 1$	$0 \leq y \leq 180$	$0 \leq x \leq \pi$
$h^{-1}(x) = \tan^{-1}(x)$ or $\arctan(x)$	Reals	$-90 < y < 90$	$\dfrac{-\pi}{2} < x < \dfrac{\pi}{2}$

While it is true that $\sin(210°) = \sin(330°) = \frac{-1}{2}$, $\sin^{-1}\left(\frac{-1}{2}\right) = -30°$ is the only acceptable answer because $y = \sin^{-1}(x)$ is a function and as such can have only one output value for a given input.

▶ Find the exact value of $\cos\left(\sin^{-1}\left(\frac{-15}{17}\right)\right)$.

▶ $\sin^{-1}\left(\frac{-15}{17}\right)$ is an angle. If you call the angle A, the equation

$\sin(A) = \frac{-15}{17}$ is equivalent (so long as A is between −90 and 0). The problem is now the equivalent of looking for the cosine of this angle.

▶ To complete the problem, find the adjacent side of a right triangle in which the opposite side is 15 and the hypotenuse is 17. Use the Pythagorean theorem to determine that the adjacent side is 8.

▶ Therefore, $\cos(A) = \frac{8}{17}$.

▶ This problem can also be solved using the Pythagorean identity $\cos^2(A) + \sin^2(A)$ and using only the positive solution to the equation because of the restriction on angle A.

Not all problems use integer Pythagorean triples and yield rational answers.

▶ Find the exact value of $\tan\left(\sin^{-1}\left(\frac{\sqrt{3}}{4}\right)\right)$.

▶ It might help if you remember $\sin^{-1}\left(\frac{\sqrt{3}}{4}\right)$ is an angle and assign that

angle some variable name. Let $B = \sin^{-1}\left(\frac{\sqrt{3}}{4}\right)$. Then $\sin(B) = \frac{\sqrt{3}}{4}$. Use the Pythagorean identity to solve $\cos^2(B)+\left(\dfrac{\sqrt{3}}{4}\right)^2=1$ to determine that $\cos(B)=\dfrac{\sqrt{13}}{4}$.

▶ Use the quotient identity to determine that $\tan(B)=\dfrac{\sqrt{3}}{\sqrt{13}}=\sqrt{\dfrac{3}{13}}$ or $\dfrac{\sqrt{39}}{13}$ if you wish to have a rational denominator.

Not only do problems need to use integer Pythagorean triples, they might contain variable expressions.

EXAMPLE

▶ Find the exact value of $\sin\left(2\tan^{-1}\left(\dfrac{x}{4}\right)\right)$.

▶ Although having the variable in the problem might make the problem look different, it is not. Let $\tan^{-1}\left(\dfrac{x}{4}\right)=C$ so that $\tan(C)=\dfrac{x}{4}$. We'll use the right triangle for this problem (although we could use the identity $1+\tan^2(C)=\sec^2(C)$. The legs of the right triangle have lengths x and 4, so the hypotenuse has length $\sqrt{x^2+16}$.

▶ We are asked to find the value of $\sin(2C)=2\sin(C)\cos(C)$. We have $\sin(C)=\dfrac{x}{\sqrt{x^2+16}}$ and $\cos(C)=\dfrac{4}{\sqrt{x^2+16}}$. We can now write

$$\sin(2C)=\sin\left(2\tan^{-1}\left(\dfrac{x}{4}\right)\right)=2\left(\dfrac{x}{\sqrt{x^2+16}}\right)\left(\dfrac{4}{\sqrt{x^2+16}}\right)=\dfrac{8x}{\sqrt{x^2+16}}.$$

▶ Observe that this is the correct answer whether x is positive or negative.

By realizing that inverse trigonometric functions are angles, we are able to manipulate a variety of problems.

EXAMPLE

▶ Find the exact value of $\cos\left(\sin^{-1}\left(\dfrac{3}{4}\right)+\tan^{-1}\left(\dfrac{4}{3}\right)\right)$.

▶ Let $A=\sin^{-1}\left(\dfrac{3}{4}\right)$ and $B=\tan^{-1}\left(\dfrac{4}{3}\right)$. This gives us $\sin(A)=\dfrac{3}{4}$ and $\tan(B)=\dfrac{4}{3}$. If $\sin(A)=\dfrac{3}{4}$, then $\cos(A)=\dfrac{\sqrt{7}}{4}$. If $\tan(B)=\dfrac{4}{3}$ then $\sin(B)=\dfrac{4}{5}$ and $\cos(B)=\dfrac{3}{5}$.

▶ The original problem can be written as $\cos(A+B)=\cos(A)\cos(B)-$

$$\sin(A)\sin(B) = \left(\dfrac{\sqrt{7}}{4}\right)\left(\dfrac{3}{5}\right)-\left(\dfrac{3}{4}\right)\left(\dfrac{4}{5}\right)=\dfrac{3\sqrt{7}-12}{20}.$$

Applying Trigonometric Functions

Ferris wheels might very well be the closest real application of the particle traveling around a circle. The riders get onto their bench and rotate to the next stop point so that the next bench can be filled, and when all the benches are filled, the ride begins.

Colin and Carson are at the amusement park and decide to ride the Ferris wheel. While waiting in line to get onto a bench on the Ferris wheel, Colin reads the following information on a board outside the control house. The wheel has a diameter of 140 feet, and it takes 60 seconds to make a complete revolution around the wheel. The center of the wheel is 80 feet above ground. Colin and Carson are the last passengers to board before the ride begins.

The height of the bench that Colin and Carson sit on can be graphed as a sinusoid with respect to the time.

▶ What is the amplitude for the sinusoid?

▶ The diameter of the wheel is 140 feet, making the height of the wheel 70 feet. The amplitude of the sinusoid for this model is 70.

One of the questions that most people ask about a Ferris wheel is how high will riders be when they reach the top of the wheel. Adding 70 feet to the 80 feet where the center of the wheel is located tells us that the boys will be 150 feet above ground when they reach the top of the Ferris wheel.

▶ Not many people consider this question: How high above the ground are the riders when the ride is about to begin?

▶ The bottom of the wheel is 70 feet below the center of the wheel. Since they are the last people to get on the Ferris wheel, they will be 10 feet off the ground when the ride begins.

It is now time for the important mathematical problem.

▶ Write an equation for the height of the bench that Colin and Carson are on as a function of time.

▶ The amplitude is 70 feet ($A = 70$), the period is 60 seconds $\left(60 = \dfrac{2\pi}{B} \Rightarrow B = \dfrac{2\pi}{60} = \dfrac{\pi}{30}\right)$, and the center of the wheel is

80 feet above the ground ($D = 80$). The equation for the height, h, of their car in terms of the number of seconds, t, into the ride is

$$h = -70\cos\left(\frac{\pi}{30}t\right) + 80.$$

▶ Colin and Carson are at the minimum height for this wheel/function at the beginning of the ride, accounting for the negative coefficient.

For many riders who are unsure of themselves about height, there is a bit of anxiety when the ride ends and they are in the air and need to wait to get off.

EXAMPLE

▶ The ride lasts 5.5 minutes. How high above ground are Colin and Carson when the ride ends? (This is before the crew goes through the motion of slowly moving the wheel from stop to stop to unload passengers.)

▶ 5.5 minutes = 330 seconds. $h(330) = 150$. Colin and Carson are at the top of the wheel, 150 feet above the ground, when the ride ends.

The Bay of Fundy is a famous tourist sight because of the difference in water depths between the low and high tides. The time difference between high and low tides is 6 hours, 13 minutes (meaning that high tides occur every 12 hours, 26 minutes). The depth of the water in the Bay of Fundy during high tide averages 43.9 feet. The depth of the water during low tides averages 6.8 feet. Assume that the depth of the water in the Bay of Fundy is a sinusoid (i.e., behaves like a sine or cosine function).

EXAMPLE

▶ Determine the amplitude for this periodic function.

▶ The amplitude is one-half the difference between the maximum and minimum values. The amplitude for this function is $\dfrac{43.9 - 6.8}{2} =$ 18.55 feet.

The average for the function is the average of the maximum and minimum values.

EXAMPLE

▶ What is the average value for this periodic function?

▶ The average depth of the water is $\dfrac{43.9 + 6.8}{2}$ = 25.35 feet.

We use the average value of the function as the vertical translation when writing an equation to model the depth of the water.

EXAMPLE

▶ Using the period between high tides to be 12.5 hours (for the ease of computation), write an equation for the depth of the water in the Bay of Fundy.

▶ Let $t = 0$ correspond to when the water is at high tide. If we begin at high tide, the graph will begin with a maximum value, so we can use a cosine function to represent the depth of the water. We know that $A = 18.55$ and $D = 25.35$.

▶ The period of the function is 12.5 hours, so $\dfrac{25}{2} = \dfrac{2\pi}{B}$. This yields $B = \dfrac{4\pi}{25}$, and the function that models the depth of the water in the Bay of Fundy is $d = 18.55\cos\left(\dfrac{4\pi}{25}t\right) + 25.35$.

Solving Trigonometric Equations

Solving trigonometric equations requires that you know the trigonometric values of the special angles, can use your calculator to determine reference angles (in both degree and radian mode), and can recognize identities.

▶ Solve $4\sin(A)+3=1$ for $0 \le A < 360°$.

▶ Subtract 3 and divide by 4 to get $\sin(A)=\dfrac{-1}{2}$. Because $\sin(30°)=\dfrac{1}{2}$, use $30°$ as the reference angle for the third and fourth quadrants.
$A = 210°, 330°$.

If the angle does not turn out to be a special angle, use your calculator to determine the reference angle and then determine the solution to the problem.

▶ Solve $8\cos(B)+4=1$ $(0 \le B < 360°)$. Answer to the nearest tenth of a degree.

▶ Subtract 4 and divide by 8 to get $\cos(B)=\dfrac{-3}{8}$. This does not represent a special angle, so the calculator will be necessary.

▶ While it is true that the calculator can compute $\cos^{-1}\left(\dfrac{-3}{8}\right)=112.0°$, you will still need to find the reference angle $(180° - 112.0° = 68.0°)$ to find the third quadrant angle $(180° + 68.0° = 248.0°)$. $B = 112.0°$, $248.0°$.

When working with quadratic forms of trigonometric equations, be alert for possible applications of trigonometric identities.

▶ Solve $4\sec^2(\theta) - 3\tan(\theta) - 5 = 0$ $(0 \le A < 360°)$. Answer to the nearest tenth of a degree.

▶ We can apply the Pythagorean identity to rewrite this equation as
$4\left(1+\tan^2(\theta)\right)-3\tan(\theta)-5=0$ and simplify this to get
$4\tan^2(\theta)-3\tan(\theta)-1=0$. Factor the quadratic to
$\left(4\tan(\theta)+1\right)\left(\tan(\theta)-1\right)=0$ and solve the equation to
get $\tan(\theta)=1,\dfrac{-1}{4}$.

▶ We know the solution to the equation $\tan(\theta)=1$ is $\theta=45°, 225°$.
To solve $\tan(\theta)=\dfrac{-1}{4}$, we need to determine the reference angle
$\tan^{-1}\left(\dfrac{1}{4}\right)=14.0°$.

▶ The solution to the equation $\tan(\theta)=\dfrac{-1}{4}$ is $166.0°$ $(180°-14.0°)$ and
$346.0°$ $(360°-14.0°)$. Therefore, the solution to the equation $4\sec^2(\theta)$
$-3\tan(\theta)-5=0$ is $\theta=45°, 166.0°, 225°, 346.0°$.

Sometimes the application of an identity creates a quadratic equation.

 IRL Periodic phenomena occur in the real world and rarely involve angles. This is
the major reason for the need for radian measure.

EXAMPLE

▶ Solve $\cos(2\theta)+3\sin(\theta)=2$ for $0\le\theta<2\pi$. Round your answer to the
nearest thousandth.

▶ There are three different identities for $\cos(\theta)$. Because the other
function in the problem is $\sin(\theta)$, use the identity $1-2\sin^2(\theta)$ for
$\cos(2\theta)$. The equation now becomes $1-2\sin^2(\theta)+3\sin(\theta)-2$ or that
$2\sin^2(\theta)-3\sin(\theta)-1=0$.

▶ Apply the quadratic formula to this equation:

$$\sin(\theta) = \frac{3 \pm \sqrt{3^2 - 4(2)(-1)}}{2(2)} = \frac{3 \pm \sqrt{17}}{4} = 1.7808, -0.2808.$$ The value

of the sine function cannot exceed 1, so we can reject 1.7808 as a value of $\sin(\theta)$. The reference angle for $\sin(\theta) = -0.2808$ is $\sin^{-1}(\theta)(0.2808) = 0.2846$.

▶ Putting this angle into the third and fourth quadrants, we get $\theta =$ 3.426 ($\pi + 0.2846$) or 6.000 ($2\pi - 0.2846$).

As you saw in an earlier section of this chapter, Colin and Carson determined that their height above ground as they rode the Ferris wheel is given by the formula $h = -70\cos\left(\dfrac{\pi}{30}t\right) + 80$.

EXAMPLE

▶ At what times during the first 150 seconds were they 125 feet off the ground?

▶ With $h = 125$, the problem becomes $125 = -70\cos\left(\dfrac{\pi}{30}t\right) + 80$. Subtract 80 and divide by –70 to get $\cos\left(\dfrac{\pi}{30}t\right) = \dfrac{-45}{70}$. A major part of this problem is to recognize that it uses radian mode. This is usually the case when working with real-life applications.

▶ The reference angle is $\cos^{-1}\left(\dfrac{45}{70}\right) = 0.8726$. We know $\cos(\theta) < 0$ when the angle is in quadrant II ($\pi - 0.8726 = 2.2690$) and quadrant III ($\pi + 0.8726 = 4.014$).

▶ Therefore, $\dfrac{\pi}{30}t = 2.2690$ when $t = 68.1$ and $\dfrac{\pi}{30}t = 4.014$ when $t = 120.4$. Colin and Carson will be 125 feet above ground 68.1 seconds after the ride starts (going up) and again 120.4 seconds after the ride starts (coming down).

The last item to consider is the idea of rotating a graph on the plane. To do this, you need to create two matrices. The first matrix will have two columns, and each row will represent the coordinates of a point on the graph of the original graph. The second matrix will be a 2 by 2 matrix of the form

$$\begin{bmatrix} \cos(\theta) & \sin(\theta) \\ -\sin(\theta) & \cos(\theta) \end{bmatrix}$$

where θ is the angle of rotation.

EXAMPLE

▶ Given the function $f(x) = x^2$, plot the points for the domain $-3 \leq x \leq x$.

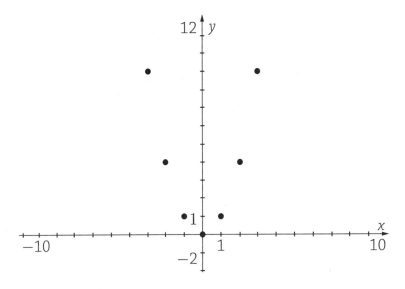

▶ When the coordinates for these points are put into a matrix and multiplied by the rotational matrix with $\theta = 45°$, the new set of coordinates are

$$\begin{bmatrix} -6\sqrt{2} & 3\sqrt{2} \\ -3\sqrt{2} & \sqrt{2} \\ -\sqrt{2} & 0 \\ 0 & 0 \\ 0 & \sqrt{2} \\ -\sqrt{2} & 3\sqrt{2} \\ -3\sqrt{2} & 6\sqrt{2} \end{bmatrix}$$

and the graph is the scatter plot of open dots shown in the diagram.

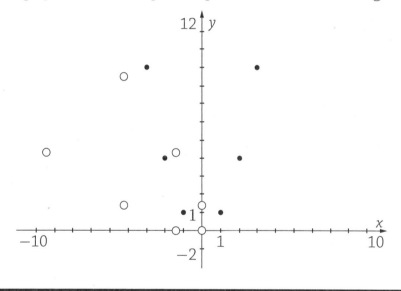

The scatter plot shows some of the points from the ellipse $\dfrac{x^2}{25} + \dfrac{y^2}{16} = 1$ after they have been rotated 60°.

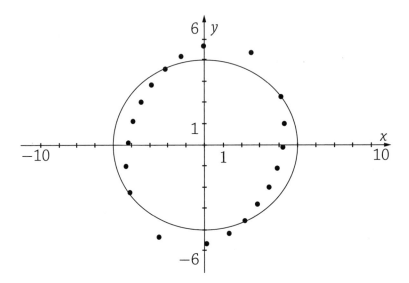

EXERCISES

EXERCISE 11-1

Describe the transformation that is applied to the base function in each problem, give the coordinates of the five key points, and then sketch a graph of this function for one period. Don't leave out question 10, though!

1. $f(\theta) = \sin(\theta) + 1$

2. $g(\theta) = \cos(\theta) + 1$

3. $f(\theta) = 2\sin(\theta)$

4. $f(\theta) = \sin(2\theta)$

5. $f(\theta) = \cos(\pi\theta)$

6. $g(\theta) = 2\cos(2\theta) - 1$

7. $k(\theta) = \dfrac{-1}{2}\sin\left(\theta - \dfrac{\pi}{2}\right)$

8. $m(\theta) = 2\sin\left(\theta + \dfrac{\pi}{4}\right) + 1$

9. $p(\theta) = -2\cos\left(\theta + \dfrac{\pi}{3}\right) + 3$

10. Given a number z with $0 \le z \le \dfrac{\pi}{2}$ and $\sin(z) = h$, what is the value of $\sin(z + 2\pi)$? Explain how you determined this answer.

EXERCISE 11-2

Write an equation for each graph.

1.

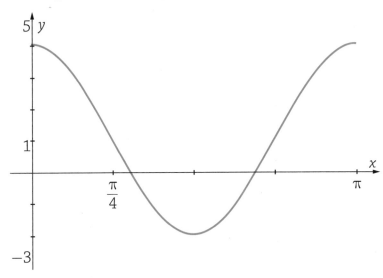

2.

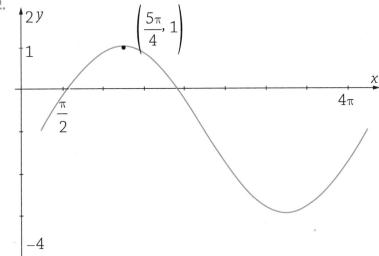

3.

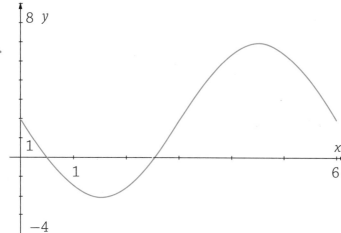

4.

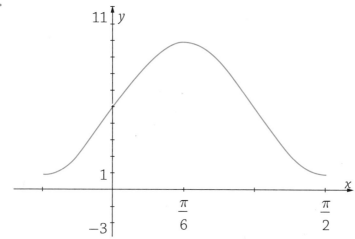

5.

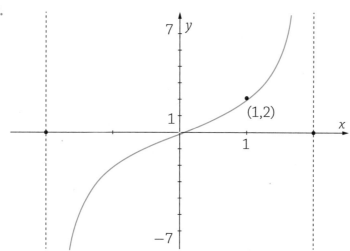

EXERCISE 11-3

For questions 1 to 4, find the exact value of each expression. Answer question 5, too!

1. $\tan\left(\sin^{-1}\left(\dfrac{-\sqrt{5}}{4}\right)\right)$

2. $\sin\left(\tan^{-1}\left(\dfrac{-5}{7}\right)\right)$

3. $\sin\left(2\tan^{-1}\left(\dfrac{3}{7}\right)\right)$

4. $\cos\left(\sin^{-1}\left(\dfrac{2}{3}\right)+\sin^{-1}\left(\dfrac{-1}{4}\right)\right)$

5. Find an algebraic expression for $\cos\left(2\tan^{-1}\left(\dfrac{x}{3}\right)\right)$

EXERCISE 11-4

Solve each of the following equations for $0 \le A \le 360°$. Round answers to the nearest tenth of a degree when necessary.

1. $4\sin(A)+3=0$

2. $5\tan(A)-9=0$

3. $2\cos^2(A)-3\cos(A)+1=0$

4. $3\sin^2(A)+5\sin(A)-2=0$

5. $7\tan^2(A)+8\tan(A)-3=0$

6. $\sin(2A)=\sin(A)$

7. $\cos(2A)+\cos(A)+1=0$

8. $2\cos(2A)-\sin(A)-1=0$

9. $4\cos(2A)-\sec(A)+4=0$

10. $\sin(2A)=\tan(A)$

EXERCISE 11-5

Solve over the domain $0\leq\theta<2\pi$. Round to the nearest hundredth where necessary.

1. $5\sin(\theta)-1=2\sin(\theta)+1$

2. $4\sin^2(2\theta)=3$

3. $3\sin^2(\theta)+\sin(\theta)-1=0$

4. $4\cos(\theta)=6\sec(\theta)-5$

5. $\csc(\theta)=2\sin(\theta)+1$

EXERCISE 11-6

Regional temperatures are predicted based on data that has been collected over a number of decades. The typical highest temperature in Rock Hill, South Carolina, 89°, occurs on July 18, while the typical lowest temperature, 41°, occurs on January 18. If we assume that these temperatures occur in a periodic manner, write the equation illustrating the typical temperature in Rock Hill. (These are typical temperatures, not any particular year. Also, as they are based on decades of data, any climate changes that have occurred in the last decade will not appear to change these values too drastically. The impact of climate change will not affect this function for another decade.) Use this information to answer the following questions,

1. What is the period for this function?

2. What is the amplitude of this function?

3. What is the average temperature for the area?

4. If $t = 0$ corresponds to January 18, write an equation to determine the high temperature in Rock Hill.

5. April 29 is 100 days after January 18. What is the predicted high temperature in Rock Hill for this day?

Flashcard App

12 Polar and Parametric Equations

escartes revolutionized the world of mathematics when he created the coordinate system because for the first time the worlds of algebra and geometry could be connected. Prior to this there were no equations for lines and there were no graphs for equations. Descartes' coordinate system is rectangular. In this chapter we will look at a coordinate system that is based on a series of concentric circles and a fixed ray from the center of those circles.

Plotting Points on a Polar Coordinate Plane

The fixed point is called the pole of the system, and the fixed ray is called the polar axis. To plot a point in the polar coordinate system, identify the radius of the circle and the angle of rotation from the pole.

EXAMPLE

▶ Plot the point (6, 30°).

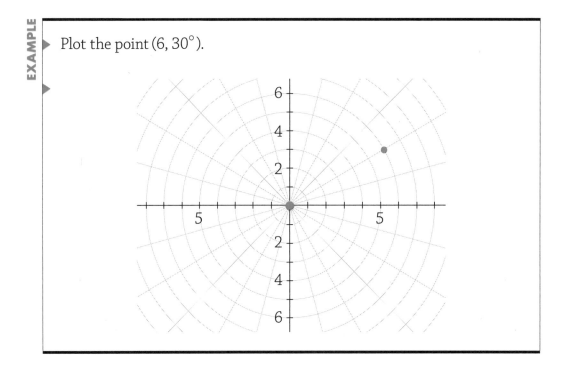

One major difference between the rectangular coordinate system and the polar coordinate system is that in rectangular coordinates there is a unique relationship between the coordinates and the point. That is to say, each point has only one set of coordinate (while each set of coordinates has only one point). In the polar coordinate system, each coordinate defines a unique point but each point has an infinite number of possible coordinates. For example, the point $(6, 30°)$ could also be $(6, 390°)$, $(6, -330°)$, $(6, 750°)$, or $(6, -690°)$. Clearly, all you need to consider is rotating around the pole a few revolutions either in a clockwise or counterclockwise rotation. However, there is another way to identify this point. The coordinates of this point could be $(-6, 210°)$. The angle $210°$ is diametrically opposed to the $30°$ angle. A positive radius indicates motion in direction of the terminal side of the angle, while a negative radius indicates motion in the reverse direction of the terminal side of the angle.

EXAMPLE

▶ Given the point $(4, 230°)$, determine a set of coordinates for the same point that has a negative radius.

▶ There are an infinite number of coordinates that will satisfy these directions, but the first set of coordinates people will likely consider is $(-4, 50°)$, since $50°$ is the reference angle for $230°$.

Here is another example for us to try.

EXAMPLE

▶ Given the point $(-8, 140°)$, determine a set of coordinates for the same point that has a positive radius.

▶ There are an infinite number of coordinates that will satisfy these directions, but the first set of coordinates people will likely consider is $(8, 320°)$, since $40°$ is the reference angle for $140°$.

Conversion Between Polar and Rectangular Coordinates

The conversion between the two coordinate systems is fairly straightforward. Given a point in the polar coordinate system with coordinates (r, θ), the rectangular coordinates are $(r\cos(\theta), r\sin(\theta))$.

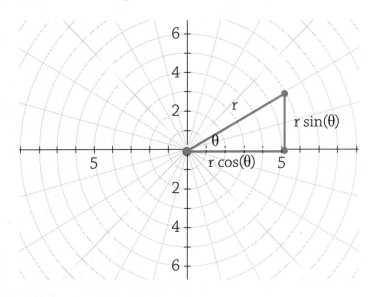

Determine the rectangular coordinates that correspond to the point with coordinates $(10, 120°)$.

$10\cos(120°) = -5$ and $10\sin(120°) = 5\sqrt{3}$, so the point in the rectangular coordinate system corresponding to $(10, 120°)$ has coordinates $(-5, 5\sqrt{3})$.

You will not always have the luxury of working with one of the special angles. Unless directed to do otherwise, rounding the answer to the nearest tenth is accurate enough for graphing purposes.

EXAMPLE

▸ Determine the rectangular coordinates that correspond to the point with coordinates $(6, 232°)$.

▸ $6\cos(232°) = -3.7$ and $6\sin(232°) = -4.7$, so the point in the rectangular coordinate system corresponding to $(6, 232°)$ has coordinates $(-3.7, -4.7)$.

A little more thought is required when you are converting coordinates from the rectangular coordinate system to the polar coordinate system. You should not be surprised that the radius is determined by the Pythagorean theorem, $r = \sqrt{x^2 + y^2}$. The angle is determined using the inverse tangent function by first determining the reference angle and then placing the angle in the appropriate quadrant.

EXAMPLE

▸ Determine one set of coordinates in the polar coordinate system that corresponds to the point $(-24, 7)$ in the rectangular coordinate system.

▸ The radius of the circle is $r = \sqrt{(-24)^2 + 7^2} = 25$. The reference angle for this set of coordinates is $\tan^{-1}\left(\dfrac{7}{24}\right) = 16.26°$. Because the point $(-24, 7)$ is in the second quadrant, the angle in question is $163.7°$ (rounded to the nearest tenth of a degree).

▸ One set of polar coordinates for the point $(-24, 7)$ is $(25, 163.7°)$.

IRL Did you ever play with the drawing toy called Spirograph? Many of the designs that can be made are based on the graphs of curves from polar coordinates!

Graphs of Polar Equations

There are seven classic curves drawn in polar coordinates that we need to examine, and then we are going to take some time to play with some variations of these. The first curve is the circle. The equation of the circle with radius a is $r = a$. Seems obvious, doesn't it? We'll look at the equation of another circle in the next section. The equation of a line in polar coordinates is $\theta = \text{angle}$. This corresponds to the equation $y = mx$ where $\theta = \tan^{-1}(m)$.

Both of these seem to be anticlimactic (though they do not answer the question about the graphs that do not contain the origin). There will be more on that later. The next graph is really interesting, $r = \theta$. This is the Spiral of Archimedes. The radius of the circle is equal to the measure of the angle. Do you have a sense of what this will look like without peeking at the graph below?

Here is the graph when θ is nonnegative.

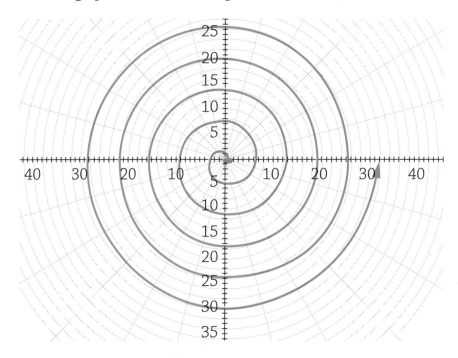

Here is the graph when θ is nonpositive.

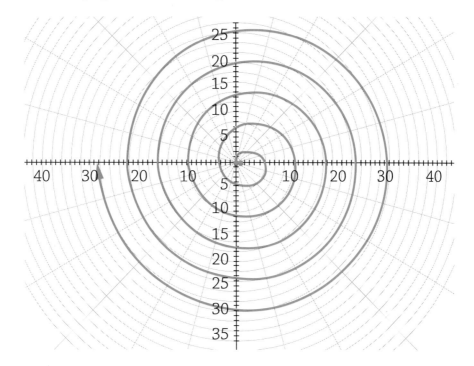

And here is the graph for when θ can be positive angles and negative angles.

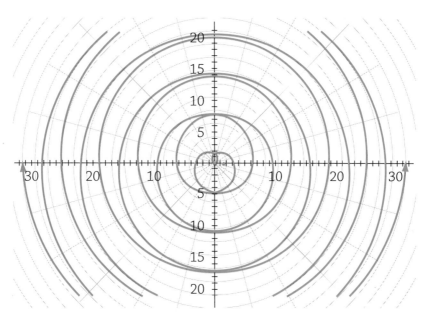

The equation can be rewritten to the form $r = a\theta + \alpha$ to get different kinds of spirals. For example, the graph of $r = 2\theta + \dfrac{\pi}{3}$ is

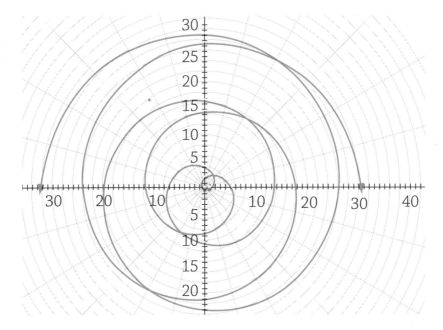

We will now take a look at the four classic curves: the rose, the cardioid, the limaçon, and the lemniscate. Each graph has a basic shape that is easily identified, and the steps for constructing the graph with pencil and paper are easily done.

The Rose

The equations for the rose are either $r = a\sin(n\theta)$ or $r = a\cos(n\theta)$, where n is an integer. The value of a dictates how long each "petal" of the rose will be. Look at the following examples of some roses, concentrating on the graphs versus the value of n.

$r = a \sin(2\theta)$

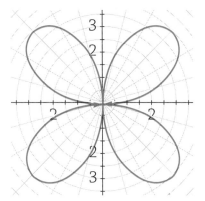

$r = a \cos(2\theta)$

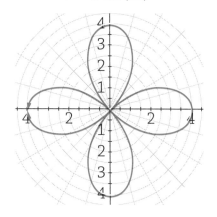

$r = a \sin(3\theta)$

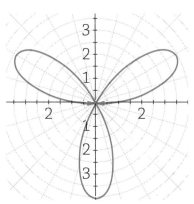

$r = a \cos(3\theta)$

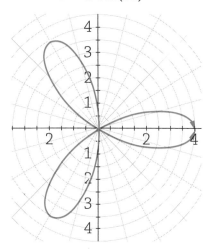

$r = a \sin(4\theta)$

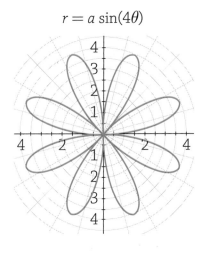

$r = a \cos(4\theta)$

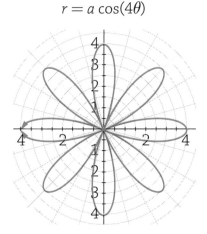

$r = a \sin(5\theta)$

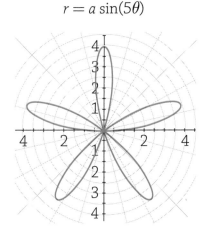

$r = a \cos(5\theta)$

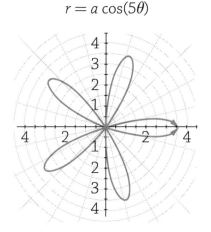

How many petals will the rose $r = a\sin(n\theta)$ or $r = a\cos(n\theta)$ have? If n is an odd number, there will be n petals, and if n is an even number, there will be $2n$ petals.

What is the angle of rotation between the tips of each petal? The angle will be $\dfrac{2\pi}{\#\text{ of petals}}$.

Given this, we are able to sketch the graph of a rose if we can sketch the first petal. For roses with the form $r = a\sin(n\theta)$, we know that the first point to sketch will be (0, 0) and the tip of the first petal will be when the value of

$n\theta = \dfrac{\pi}{2}$ or $\theta = \dfrac{\pi}{2n}$. Therefore, the coordinates of the tip of the first petal will

be $\left(a, \dfrac{\pi}{2n}\right)$. For roses with the form $r = a\cos(n\theta)$, the tip of the first petal

will be $(a, 0)$ and the graph will cross through the pole when $\theta = \dfrac{\pi}{2n}$. We can

use the symmetry of each petal to draw the rest of the first petal and then use rotational symmetry to finish the graph.

EXAMPLE

▶ Determine the coordinates of the tip of each petal of the sketch of $r = 6\sin(8\theta)$.

▶ $8\theta = \dfrac{\pi}{2}$ when $\theta = \dfrac{\pi}{16}$, so the coordinates of the tip of the first petal

will be $\left(8, \dfrac{\pi}{16}\right)$. Since there are 16 petals in the graph of $r = 6\sin(8\theta)$,

the remaining petals are at an angle of $\dfrac{2\pi}{16} = \dfrac{\pi}{8}$ from each other.

▶ The coordinates are $\left(8, \dfrac{3\pi}{16}\right)$, $\left(8, \dfrac{5\pi}{16}\right)$, $\left(8, \dfrac{7\pi}{16}\right)$, $\left(8, \dfrac{9\pi}{16}\right)$, $\left(8, \dfrac{11\pi}{16}\right)$,

$\left(8, \dfrac{13\pi}{16}\right)$, $\left(8, \dfrac{15\pi}{16}\right)$, $\left(8, \dfrac{17\pi}{16}\right)$, $\left(8, \dfrac{19\pi}{16}\right)$, $\left(8, \dfrac{21\pi}{16}\right)$, $\left(8, \dfrac{23\pi}{16}\right)$, $\left(8, \dfrac{25\pi}{16}\right)$,

$\left(8, \dfrac{27\pi}{16}\right)$, $\left(8, \dfrac{29\pi}{16}\right)$, and $\left(8, \dfrac{31\pi}{16}\right)$.

We know how to find the tip of the first petal when the equation of the graph is of the form $r = a\cos(n\theta)$. Once we realize how easy it is to sketch, the graph is fun to do.

Let's sketch the graph of $r = 2\cos(3\theta)$. The graph begins at the point $(2, 0)$. The sketch of the graph on the interval $0 \le \theta \le \dfrac{\pi}{6}$ is:

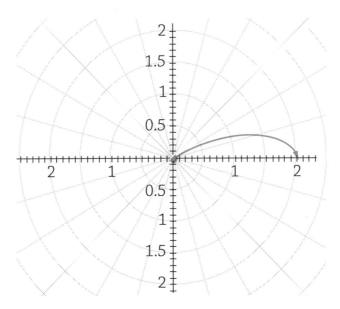

The sketch of the graph on the interval $0 \leq \theta \leq \dfrac{\pi}{3}$ is:

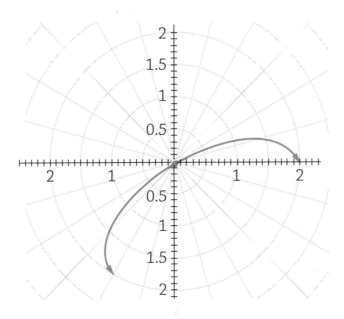

Continuing on to the interval $0 \le \theta \le \dfrac{2\pi}{3}$, the sketch is:

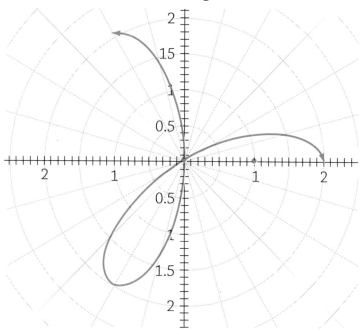

Finally, the sketch on the interval $0 \le \theta \le \pi$ completes the graph.

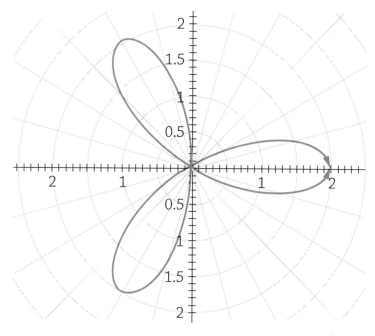

What happens to the graph on the interval $\pi \leq \theta \leq 2\pi$? The graph traces over itself.

EXAMPLE

▶ Sketch the graph of $r = 3\cos(2\theta)$.

▶ The tip of the first petal is at $(3, 0)$, and the graph will pass through the pole at $\left(0, \dfrac{\pi}{4}\right)$. (Remember, there are four petals, so we need to compute the value $\dfrac{\pi}{2(2)}$.)

▶ The sketch of the graph on the interval $0 \leq \theta \leq \dfrac{\pi}{4}$ is:

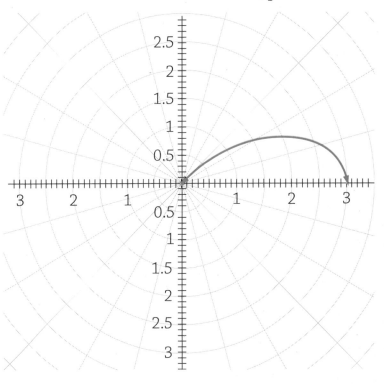

▶ The sketch of the graph on the interval $0 \le \theta \le \dfrac{\pi}{2}$ is:

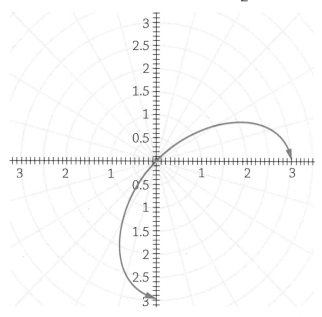

▶ The sketch of the graph on the interval $0 \le \theta \le \pi$ is:

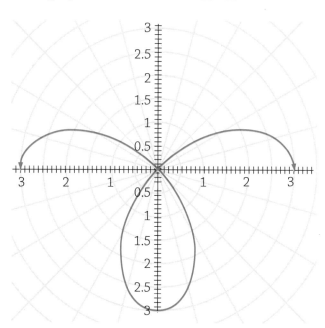

▶ Do you see that we are halfway through the interval $0 \leq \theta \leq 2\pi$ and only half the picture is complete? As you can see, the graph of the function is:

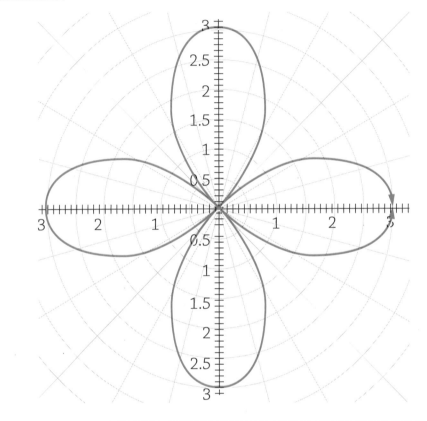

The Cardioid

Even if you did not know anything about functions in polar coordinates, the name alone gives you a clue as to how this function will look. The form of the equation for the cardioid is $r = a \pm a\sin(\theta)$ or $r = a \pm a\cos(\theta)$.

The largest value of r will be $2a$, and the smallest value of r will be 0. These will occur when the values of the trigonometric functions are ± 1.

Sketch the graph of $r = 2 - 2\sin(\theta)$.

We know when $\sin(\theta) = 0$, $r = 2$; when $\sin(\theta) = 1$, $r = 0$; when

$\sin(\theta) = -1$, $r = 2$. The graph begins at $(2, 0)$, "rises" for a while to

return to $\left(0, \dfrac{\pi}{2}\right)$ (remember, when $\theta = \dfrac{\pi}{6}$, $r = 1$), "rises" again and then

falls to $(2, \pi)$, continues to "fall" to $\left(2, \dfrac{3\pi}{2}\right)$, and ends at $(2, 2\pi)$.

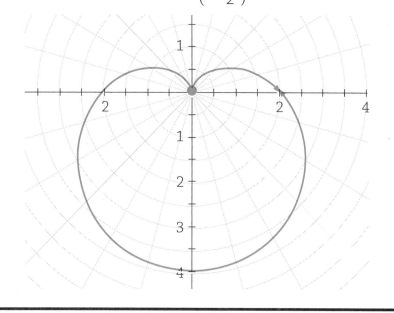

The variations of this basic graph using addition or subtraction and either the sine or cosine function cause the cardioid to rotate 90° about the plane. Try them so that you can get a feel for them.

The Limaçon

The equation for the limaçon looks like the equation for the cardioid with the exception that the coefficients do not have to be the same. That is, the equations for the limaçon are $r = a \pm b\sin(\theta)$ or $r = a \pm b\cos(\theta)$. The maximum value for r is $a + b$, while the minimum value will be the difference between a and b. Let's explore this with a few examples.

▶ Analyze and sketch the graph of $r = 5 + 3\cos(\theta)$.

▶ The largest value of the function is 8 and occurs at (8, 0), while the smallest value is 2 at the point (2, π). In addition, when $\cos(\theta) = 0, r = 5$.

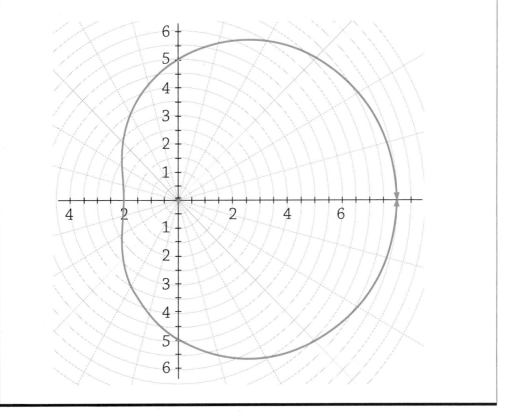

This next example is similar.

EXAMPLE

▶ Analyze and sketch the graph of $r = 3 - 2\sin(\theta)$.

▶ The largest value of the function is 5 and occurs at $\left(5, \dfrac{3\pi}{2}\right)$, while the smallest value is 1 at the point $\left(1, \dfrac{\pi}{2}\right)$. When $\sin(\theta) = 0$, $r = 3$.

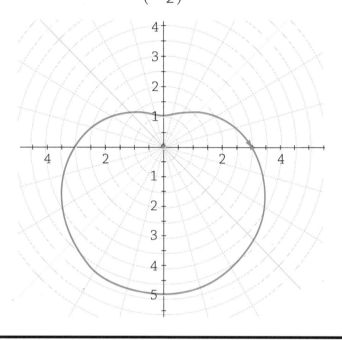

Things get much more interesting when $|b| > |a|$.

EXAMPLE

▶ Analyze and sketch the graph of $r = 2 + 4\cos(\theta)$.

▶ The largest value of the function is 6 and occurs at $(6, 0)$, while the smallest value is −2 at the point $(-2, \pi)$. When $\cos(\theta) = 0$, $r = 2$. If the smallest value of r is negative, there must be at least one place when the value of r is zero. Solve the equation $2 + 4\cos(\theta) = 0$ to get $\cos(\theta) = \dfrac{-1}{2}$, which yields $\theta = \dfrac{2\pi}{3}, \dfrac{4\pi}{3}$.

▶ Let's sketch the graph in stages. The graph of the function on the interval $0 \le \theta \le \dfrac{2\pi}{3}$ is:

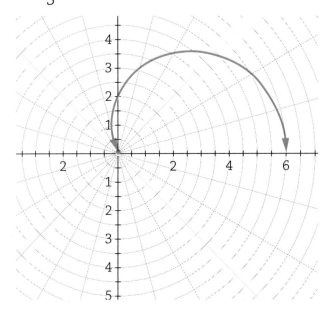

▶ The graph of the function on the interval $0 \le \theta \le \dfrac{4\pi}{3}$ is:

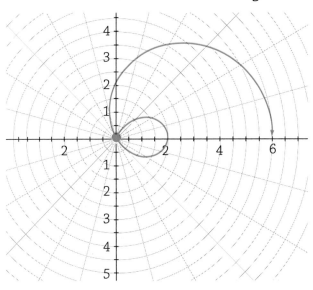

▸ The interval from $\dfrac{2\pi}{3} \le \theta \le \dfrac{4\pi}{3}$ forms the inner loop of the limaçon.

▸ We finish the graph for the interval $0 \le \theta \le 2\pi$:

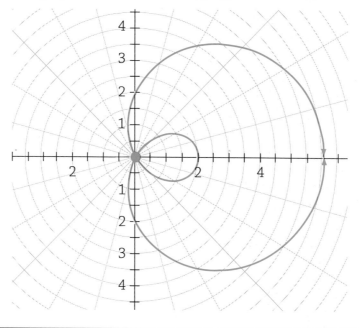

As you should well understand, the value of the angle for when the graph passes through the pole will not always be one of the special angles. As we did in the last chapter, we'll use our calculator and get estimates for the values.

▸ Analyze and sketch the graph of $r = 3 + 4\sin(\theta)$.

▸ The largest value of the function is 8 and occurs at $\left(7, \dfrac{\pi}{2}\right)$, while the smallest value is –1 at the point $\left(-1, \dfrac{3\pi}{2}\right)$. When $\sin(\theta) = 0, r = 3$.

If the smallest value of r is negative, there must be at least one place when the value of r is zero.

▶ Solve the equation $3+4\sin(\theta)=0$ to get $\sin(\theta)=\dfrac{-3}{4}$. The reference angle for the solution is $\alpha=0.8481$, so the solutions to this equation will be $\theta=3.99$, $(\pi+\alpha)$, and $\theta=5.44$, $(2\pi-\alpha)$. The sketch of the graph is

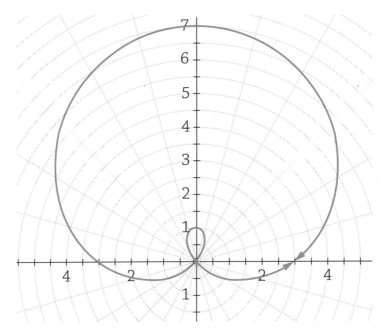

The Lemniscate

The last of the classic curves we will look at is the lemniscate. The equation for the lemniscate is $r^2=2a^2\sin(2\theta)$, or $r^2=2a^2\cos(2\theta)$. Its definition is similar to that of the ellipse. Given two foci that are $2a$ units apart, a point P is on the lemniscate if the product $PF_1\times PF_2=a^2$.

The graph will be symmetric about the pole since $r=\pm a\sqrt{2}\sqrt{\sin(2\theta)}$ or $r=\pm a\sqrt{2}\sqrt{\cos(2\theta)}$.

EXAMPLE

▶ Analyze and sketch the graph of $r^2 = 8\sin(2\theta)$.

▶ When $\sin(2\theta) = 1$, $r = \pm 2\sqrt{2}$, which represent the minimum and maximum values of the function. The graph will pass through the poles when $\sin(2\theta) = 0$.

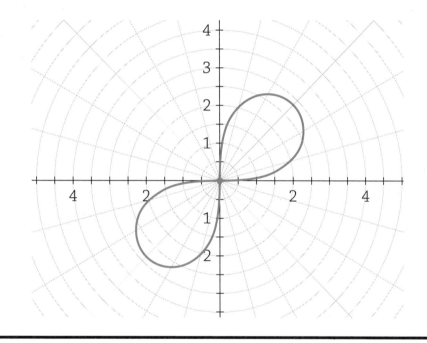

The lemniscate defined from the cosine function will also be symmetric about the polar axis.

Conic Sections in Polar Coordinates

The eccentricity, *e,* of the conic section is a key piece to writing the equation of the conic section in polar coordinates. Recall if *e* = 1, the conic is a parabola. The conic is an ellipse if *e* < 1 and a hyperbola if *e* > 1.

The basic equation for the conics is $r=\dfrac{ep}{1\pm e\sin(\theta)}$ or $r=\dfrac{ep}{1\pm e\cos(\theta)}$.
The variable p indicates the location of the conic's directrix. If the denominator contains the sine function, the directrix is the horizontal line $y = p$, and if the denominator contains the cosine function, the directrix is the vertical line $x = p$.

▶ Identify the type of conic represented by the equation $r=\dfrac{12}{4+3\cos(\theta)}$, identify some key points for the graph, and sketch a graph.

▶ The key to this process is the number 1 in the denominator of the fraction. Divide the numerator and denominator by 4.

$$r=\dfrac{3}{1+\dfrac{3}{4}\cos(\theta)}$$

▶ The coefficient of the cosine function tells us that the eccentricity is $\dfrac{3}{4}$, making the conic an ellipse. We can also rewrite the numerator to be

$$r=\dfrac{\left(\dfrac{3}{4}\right)4}{1+\dfrac{3}{4}\cos(\theta)}$$

▶ We now know that the line $x = 4$ is the directrix for the curve.

▶ We use the values $\theta=0, \dfrac{\pi}{2}, \pi$, and $\dfrac{3\pi}{2}$ as the key points to sketch the ellipse. Using these values, we get the points $\left(\dfrac{12}{7},0\right)$, $\left(3,\dfrac{\pi}{2}\right)$, $(12,\pi)$, and $\left(3,\dfrac{3\pi}{2}\right)$.

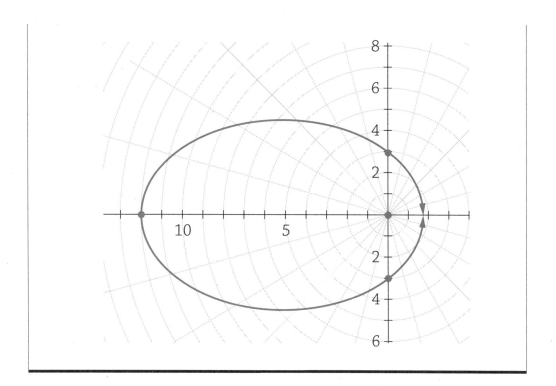

This next example uses a sine function.

▶ Identify the type of conic represented by the equation $r = \dfrac{10}{3 - 5\sin(\theta)}$, identify some key points for the graph, and sketch a graph.

▶ Divide the numerator and denominator by 3, and adjust the numerator to include the value of e.

$$r = \dfrac{\left(\dfrac{5}{3}\right)2}{1 - \dfrac{5}{3}\sin(\theta)}$$

▶ The eccentricity is $\dfrac{5}{3}$ telling us that the conic is a hyperbola, and the value of $p = 2$ in the numerator indicates that the directrix is the line $y = 2$.

▶ The key points for this graph are $\left(\dfrac{10}{3}, 0\right)$, $\left(-5, \dfrac{\pi}{2}\right)$, $\left(\dfrac{10}{3}, \pi\right)$, and $\left(\dfrac{5}{4}, \dfrac{3\pi}{2}\right)$.

▶ The sketch of the graph is

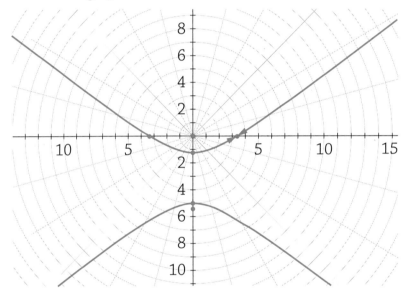

Nonstandard Polar Curves

Having access to graphing technology gives us the opportunity to play with the polar functions. I will limit the discussion (mostly) to variations of the classic curves. I strongly encourage you to play with your creations. You never know what you will find.

The equation for the rose requires that n be an integer. Try using a decimal value for the value of n.

EXAMPLE

▶ Use your graphing technology to sketch the graph of $r=4\cos(2.1\theta)$.

▶ The graph for $r=4\cos(2.1\theta)$ on the interval $0\leq\theta\leq2\pi$ doesn't seem all that impressive.

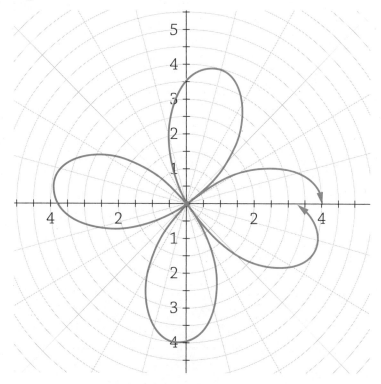

▶ However, if you change the domain to $0\leq\theta\leq10\pi$, you get a much more interesting picture.

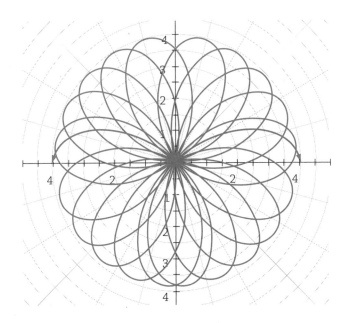

▶ Even better, change the domain to $0 \leq \theta \leq 20\pi$.

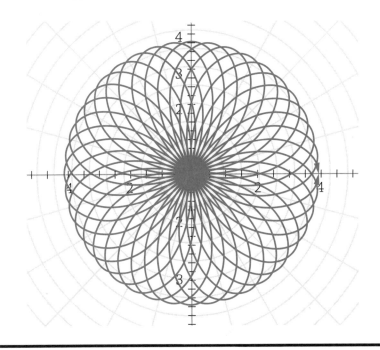

The cardioid and the limaçon have the form $r=a+b\sin(\theta)$ or $r=a+b\cos(\theta)$. What if the coefficient of θ is not 1? Let's look at a couple of examples using the domain $0\leq\theta\leq 20\pi$.

▶ Use your graphing technology to sketch the graph of $r=3+4\cos(2\theta)$.

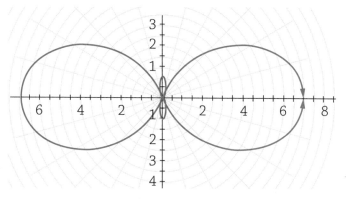

▶ That's not bad, but it's nowhere near as impressive as the last graph. What happens if we make the coefficient 2.1 rather than 2 and $r=3+4\cos(2.1\theta)$?

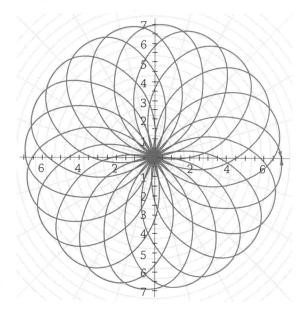

▶ That's much better. You're thinking about what would happen if we extended the domain, aren't you? The sketch of the graph for the domain $0 \leq \theta \leq 80\pi$ is really interesting.

▶ We are pushing the graphing device to the end of its tolerance, so we start to see some sharp edges. It does add something to the picture!

You can play with the graphs using more than one decimal place, using irrational numbers as coefficients, or even using "interesting fractions" (such as a denominator with a large prime denominator). Hope you enjoy playing. Who said math can't be fun?

EXAMPLE

▶ Use your graphing technology to sketch the graph of
$r = 3 + \sin(2\theta) + \cos(3\theta)$.

▶ This is a bit different than the other examples. While it is not as
complex as the previous graphs, it does seem to be rather interesting.

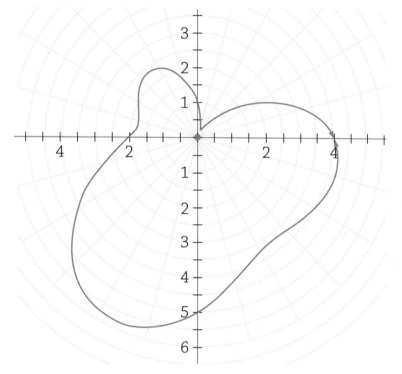

▶ Do you see a mitten?

Conversion Between Polar and Rectangular Forms for a Function

We use the conversion factors $x = r\cos(\theta)$ and $y = r\sin(\theta)$ to convert
equations from rectangular format to polar coordinate format.

▶ Write the equation for the line $4x+3y=5$ as a polar coordinate equation.

▶ Use the conversion factor to create the equation

$4r\cos(\theta)+3r\sin(\theta)=5$. Rewriting this equation in the form $r =$ we get

$$r=\frac{5}{4\cos(\theta)+3\sin(\theta)}.$$

That's not too exciting, but it works. Let's try it the other way.

▶ Write the equation $r = 4 \cos (\theta)$ in rectangular form.

▶ Multiply both sides of the equation by r so that the right-hand side of the equation will contain the expression $r\cos(\theta)$. The equation becomes $r^2=4r\cos(\theta)$. We know $r^2=x^2+ y^2$, so the equation becomes $x^2+y^2=4x$.

▶ When modified, this equation becomes
$x^2 - 4x +y^2=0 \Rightarrow x^2 - 4x + 4 +y^2=4 \Rightarrow (x-2)^2+y^2=4$, circle with center (2, 0) and radius equal to 2.

This next example is a bit trickier.

▶ Write the equation for the rose $r=4\sin(2\theta)$ in rectangular form.

▶ The critical issue here is that the conversion factor is $y=r\sin(\theta)$, not $y=r\sin(2\theta)$. Use the trigonometric identity for sin (2θ) to change the equation $r=4\sin(2\theta)$ to $r= 8\sin (\theta) \cos (\theta)$. Each of the trigonometric functions needs a factor of r in order for the conversion to work, so multiply both sides of the equation by r^2. The equation $r=8\sin(\theta)\cos(\theta)$ now becomes $r^3=8(r\sin(\theta))(r\cos(\theta))$.

▶ Converting this equation to rectangular form, we get $\left(x^2+y^2\right)^{\frac{3}{2}}=8xy$. Now you know why you never studied the roses in math class prior to polar coordinates.

We looked at the rose. We should also look at the cardioid.

▶ Write the equation for the cardioid $r=-1+2\sin(\theta)$ in rectangular form.

▶ Multiply both sides of the equation by r. The equation

$r=-1+2\sin(\theta)$ becomes $r^2=-r+2r\sin(\theta)$, and this converts to

$x^2+y^2=-\sqrt{x^2+y^2}+2y$.

▶ This is another equation that does not play well in rectangular coordinates.

Let's look at one last example, the equation of the parabola.

▶ Write the equation $y=x^2$ in polar coordinate form.

▶ Use the conversion factors to change $y=x^2$ to

$r\sin(\theta)=\left(r\cos(\theta)\right)^2=r^2\cos^2(\theta)$. Divide both sides of the equation by

the product $r\cos^2(\theta)$ to get $r=\dfrac{\sin(\theta)}{\cos^2(\theta)}=\tan(\theta)\sec(\theta)$.

Conversion Between Parametric and Function Forms

Parametric equations are very useful to model situations in two or three dimensions when the independent variable is not one of the dimensions. That is to say, we normally write functions in the form $y=f(x)$. We use parametric equations when the reality is $x=g(t)$ and $y=k(t)$.

There will be plenty of times when these two equations can be combined into a function of the form $y=f(x)$ and the third parameter will be hidden in the background. There are other scenarios when this cannot happen and we leave the problem in terms of both parametric equations.

EXAMPLE

▶ Rewrite the parametric equations $x(t)=\sqrt{t}$ and $y(t)=t$ as a single function of y in terms of x.

▶ If $x=\sqrt{t}$ then $x^2=t$. Substitute for t in the second equation to get $y=x^2$.

▶ What is the difference between $x(t)=\sqrt{t}$ and $y(t)=t$ and the traditional $y=x^2$? The domain of $y=x^2$ is traditionally the set of real numbers. However, with the parametric equations, the domain is $t>0$. The graph of the parametric equation only graphs the right-hand side of the parabola.

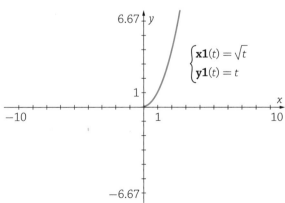

$$\begin{cases} \mathbf{x1}(t)=\sqrt{t} \\ \mathbf{y1}(t)=t \end{cases}$$

There will be times when the third parameter will be the angle θ rather than time.

EXAMPLE

▶ The motion of a particle is given by the formula $x(\theta)=4\cos(\theta)$ and $y(\theta)=5\sin(\theta)$. Describe the motion of the particle in terms of x and y.

▶ We will rewrite the two defining functions as $\dfrac{x}{4}=\cos(\theta)$ and $\dfrac{y}{5}=\sin(\theta)$. We can now use the Pythagorean identity

$\cos^2(\theta)+\sin^2(\theta)=1$ to derive the desired equation, $\dfrac{x^2}{16}+\dfrac{y^2}{25}=1$. The

particle travels in an elliptical path.

As mentioned at the beginning of this section, there are times when the model cannot, or should not, be written in a single equation.

Suppose that Jackie hit a golf ball with an initial velocity of 150 feet per second (or 102 mph) at an angle of 30° to the horizontal off of a tee box that is elevated 10 feet. We'd like to write a model to describe the location of the ball t seconds after the ball has been hit. A vector diagram of the ball being struck off the tee shows

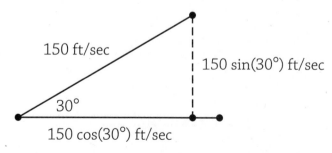

that the ball moves vertically at a speed of $150\sin(30°)=75$ feet per second and horizontally at a speed of $150\cos(30°)=75\sqrt{3}$ feet per second.

▶ Find parametric equations that describe the position of the ball as a function of time.

▶ The horizontal position of the ball is strictly determined by the force supplied by the golfer when the ball is struck. In this case, $x(t) = 75\sqrt{3}\,t$. The vertical position of the ball is determined by the initial height of the ball (the tee box is elevated 10 feet above the fairway and we ignore the height of the tee), the force supplied by the golfer (75 feet per second), and gravity (–16 feet per second per second).

▶ Yes, we do ignore frictional resistance, and we get to assume that this is a perfect day with no wind impacting the ball!

▶ The vertical position of the ball is given by the equation $y(t) = -16t^2 + 75t + 10$.

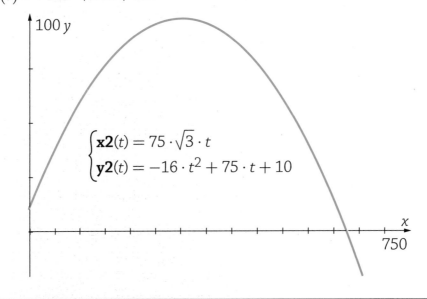

$$\begin{cases} \mathbf{x2}(t) = 75 \cdot \sqrt{3} \cdot t \\ \mathbf{y2}(t) = -16 \cdot t^2 + 75 \cdot t + 10 \end{cases}$$

As you can see, this model needs to be modified a little bit since when the ball lands on the ground, it will (hopefully) bounce forward and not bury itself in the ground. This brings us to the next question.

EXAMPLE

How long is the golf ball in the air?

The ball is in the air when $t > 0$ until $y = 0$. Solve the equation $-16t^2 + 75t + 10 = 0$ to determine when the ball hits the ground.

$$t = \frac{-75 \pm \sqrt{75^2 - 4(-16)(10)}}{-32} = -0.12, 4.82 \text{ seconds}$$

We reject the negative answer because the ball has yet to be struck. The ball strikes the ground approximately 4.82 seconds after it was hit. Modifying the domain to reflect this, the graph now looks like this.

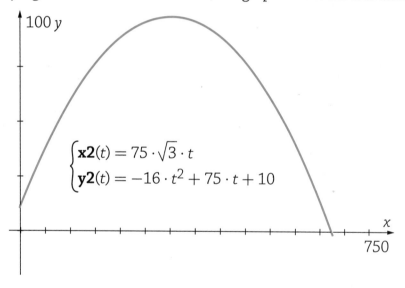

$$\begin{cases} x2(t) = 75 \cdot \sqrt{3} \cdot t \\ y2(t) = -16 \cdot t^2 + 75 \cdot t + 10 \end{cases}$$

Jackie hit that ball pretty well. Good for her. Let's try another one.

EXAMPLE

Determine the maximum height of the ball.

This is just a matter of finding the vertex of the parabola. Unfortunately, the maximum feature does not work on the calculator when in parametric mode because the independent variable, t, is not being graphed.

▶ We go back to the algebraic process to determine the ball reached its maximum height when $t = \dfrac{-75}{-32} = 2.34$ seconds. The height of the ball at this point is 97.9 feet.

As far as Jackie is concerned, that is all well and good, but she wants to know how far the ball went so she can plan her next shot. Let's be fair to Jackie and argue that once the ball landed, it bounced and rolled another 20 yards toward the pin.

▶ Determine the distance that the ball traveled.

▶ The ball is 625.8 feet ($x(4.82)$) from the tee box when it lands. Add another 60 feet for the roll, and the ball is 685.8 feet (228.6 yards) from the tee box when it stops rolling.

EXERCISES

EXERCISE 12-1

For each question, find a set of coordinates for the given point using a negative radius.

1. $\left(5, 95°\right)$

2. $\left(6, 215°\right)$

3. $\left(7, \dfrac{7\pi}{6}\right)$

4. $\left(8, \dfrac{-8\pi}{9}\right)$

EXERCISE 12-2

Convert each of the coordinates from rectangular coordinates to polar coordinates.

1. $\left(3\sqrt{2}, 3\sqrt{2}\right)$ (angle in degrees)

2. $\left(-5\sqrt{3}, -5\right)$ (angle in radians)

3. $\left(-4, 5\right)$ (angle in radians)

EXERCISE 12-3

Convert each of the coordinates from polar coordinates to rectangular coordinates.

1. $\left(12, 210°\right)$

2. $\left(10, \dfrac{3\pi}{4}\right)$

3. $\left(-20, \dfrac{3\pi}{5}\right)$

EXERCISE 12-4

Determine the coordinates of the tips of the petals for the roses defined in each question.

1. $r=4\cos(3\theta)$

2. $r=8\sin(5\theta)$

3. $r=10\cos(6\theta)$

4. $r=5\sin(4\theta)$

EXERCISE 12-5

Determine the coordinates of the key points of the cardioids defined in each of the following questions.

1. $r=5+5\cos(\theta)$

2. $r=4+4\sin(\theta)$

3. $r=3-3\cos(\theta)$

4. $r=6-6\sin(\theta)$

EXERCISE 12-6

Determine the coordinates of the key points of the limaçons defined in each question. Include the coordinates of the points where the limaçons cross the pole. (Answers in radian values.)

1. $r=3+6\cos(\theta)$

2. $r=2\sqrt{2}+4\sin(\theta)$

3. $r=3-5\cos(\theta)$

4. $r=4-6\sin(\theta)$

EXERCISE 12-7

Determine the equation for each of the following polar graphs.

1.

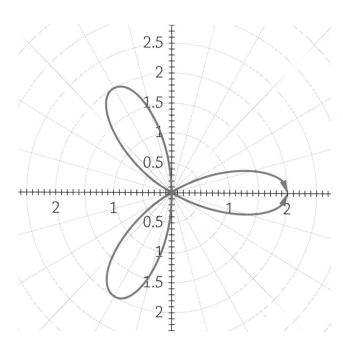

2.

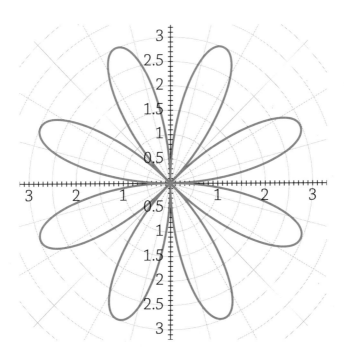

3.

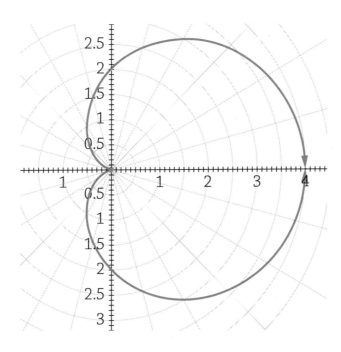

4.

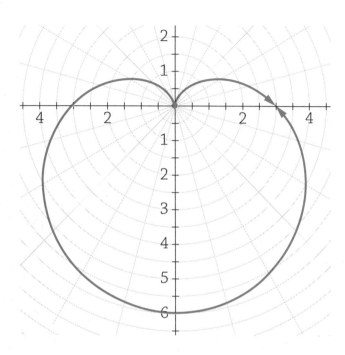

5.

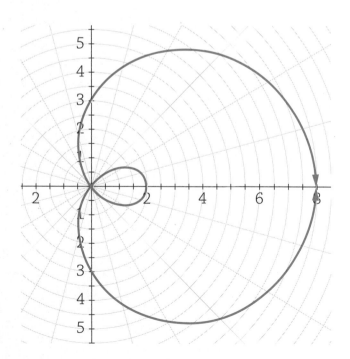

6.

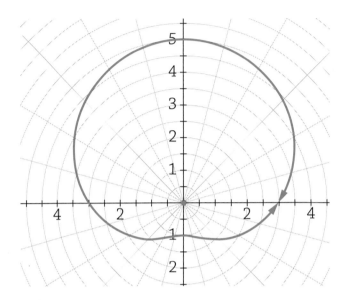

7.

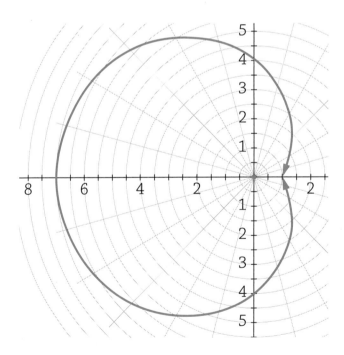

8.

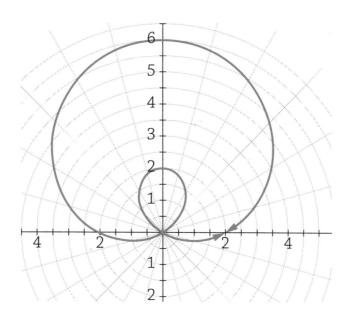

EXERCISE 12-8

Identify the conic whose equation is given in each question.

1. $r = \dfrac{15}{5 + 3\cos(\theta)}$

2. $r = \dfrac{15}{3 - 5\cos(\theta)}$

3. $r = \dfrac{15}{5 + 5\cos(\theta)}$

EXERCISE 12-9

For each question, rewrite each pair of parametric equation as a single function of the form $y = f(x)$.

1. $x(t) = 2t - 3;\ y(t) = \dfrac{-2}{3}t + 1$

2. $x(t) = 2^t;\ y(t) = 2t + 3$

3. $x(t)=4\sec(t); y(t)=3\tan(t)$

4. $x(t)=6\sin(t); y(t)=3\cos(t)$

EXERCISE 12-10

A ball is thrown in the air with a speed of 81.5 feet per second at an angle of 40° from the top of a building 50 feet above the ground:

1. Write the parametric equations to model this situation.

2. What is the maximum height of the ball?

3. How far from the base of the building does the ball strike the ground?

Flashcard
App

13 Complex Numbers

he Fundamental Theorem of Algebra tells us that there are n roots from the set of complex numbers that solve the equation $z^n = a$. We have the ability to easily find these roots when $n = 2$, 3, or 4 using factoring techniques. Beyond that, it is a very laborious process. We'll find that it is much easier to find these roots when the numbers are written in polar form. We'll also find that it is much easier to multiply and divide complex numbers in polar form than in rectangular form (but not so much for addition and subtraction).

Polar Form of Complex Numbers

Because the rectangular representation of a complex number is the ordered pair (a, b), where a represents the real portion of the number and b represents the imaginary part of the complex number, the technique for converting complex numbers into polar form from rectangular form is really the same as for any ordered pair. That is to say:

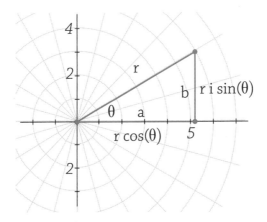

$$a + bi = r\cos(\theta) + r\sin(\theta)i$$

The abbreviation for $r\cos(\theta) + r\sin(\theta)i$ is $r\,cis(\theta)$.

EXAMPLE

▶ Convert $10\,cis\left(\dfrac{\pi}{3}\right)$ to rectangular form.

▶ $10\,cis\left(\dfrac{\pi}{3}\right) = 10\cos\left(\dfrac{\pi}{3}\right) + 10i\sin\left(\dfrac{\pi}{3}\right) = 10\left(\dfrac{1}{2}\right) + 10i\left(\dfrac{\sqrt{3}}{2}\right) = 5 + 5i\sqrt{3}$

We still need to be aware that when either a or b is a negative number, we may need to use a reference angle to find the correct value of the angle (with the exception being if the angle is one of the special angles).

EXAMPLE

▶ Convert $8 - 8i\sqrt{3}$ to polar form.

▶ The value of r is $\sqrt{8^2 + \left(8\sqrt{3}\right)^2} = 16$ while $\theta = \dfrac{5\pi}{3}$.

Therefore, $8 - 8i\sqrt{3} = 16\,cis\left(\dfrac{5\pi}{3}\right)$.

Not all problems are so straightforward. Here is an example.

EXAMPLE

▶ Convert $-3 + 4i$ to polar form.

▶ You recognize the 3–4–5 triangle and thus know that $r = 5$.

Use the reference angle $\tan^{-1}\left(\dfrac{3}{4}\right) = 0.644$ to determine that $\theta = \pi - 0.644 = 2.50$ (rounded to the nearest hundredth).

▶ Therefore, $-3 + 4i = 5\,cis(2.50)$.

Multiplication and Division of Complex Numbers

The product of the two complex numbers $r_1\, cis(\alpha)$ and $r_2\, cis(\beta)$ is $r_1 r_2\, cis(\alpha + \beta)$. Observe how the trigonometric identities for the cosine and sine of the sum of two angles work in this product.

$$(r_1\, cis(\alpha))(r_2\, cis(\beta)) = (r_1 \cos(\alpha) + r_1 i\sin(\alpha))(r_2 \cos(\beta) + r_2 i\sin(\beta))$$

- Expand the product.

$$(r_1\, cis(\alpha))(r_2\, cis(\beta)) = r_1 r_2 \cos(\alpha)\cos(\beta) + r_1 r_2 i\cos(\alpha)\sin(\beta)$$
$$+ r_1 r_2 i\sin(\alpha)\cos(\beta) + r_1 r_2 i^2 \sin(\alpha)\sin(\beta)$$

- Combine real and imaginary terms.

$$(r_1\, cis(\alpha))(r_2\, cis(\beta)) = r_1 r_2 \cos(\alpha)\cos(\beta) - r_1 r_2 \sin(\alpha)\sin(\beta)$$
$$+ (r_1 r_2 \cos(\alpha)\sin(\beta) + r_1 r_2 \sin(\alpha)\cos(\beta))i$$

- Factor $r_1 r_2$ from the terms.

$$(r_1\, cis(\alpha))(r_2\, cis(\beta)) = r_1 r_2 \big[\cos(\alpha)\cos(\beta) - \sin(\alpha)\sin(\beta)$$
$$+ (\cos(\alpha)\sin(\beta) + \sin(\alpha)\cos(\beta))i\big]$$

$$\cos(\alpha)\cos(\beta) - \sin(\alpha)\sin(\beta) = \cos(\alpha + \beta)\, \text{and}$$
$$\cos(\alpha)\sin(\beta) + \sin(\alpha)\cos(\beta) = \sin(\alpha + \beta)$$

- Substituting these sum identities into the expansion, we get the desired result.

$$(r_1\, cis(\alpha))(r_2\, cis(\beta)) = r_1 r_2[\cos(\alpha + \beta) + i\sin(\alpha + \beta)] = r_1 r_2 cis(\alpha + \beta)$$

EXAMPLE

▸ Find the product of $z_1 = 12cis\left(\dfrac{2\pi}{3}\right)$ and $z_2 = 10cis\left(\dfrac{3\pi}{4}\right)$.

▸ $z_1 z_2 = (12)(10)cis\left(\dfrac{2\pi}{3} + \dfrac{3\pi}{4}\right) = 120cis\left(\dfrac{17\pi}{12}\right)$

Compare the ease of finding the product in polar form to finding the product of $z_1 = -6 + 6i\sqrt{3}$ and $z_2 = -5\sqrt{2} + 5i\sqrt{2}$.

$$z_1 z_2 = \left(-6 + 6i\sqrt{3}\right)\left(-5\sqrt{2} + 5i\sqrt{2}\right) = 30\sqrt{2} - 30i\sqrt{2} - 30i\sqrt{6}$$
$$+ 30i^2\sqrt{6} = \left(30\sqrt{2} - 30\sqrt{6}\right) + \left(-30\sqrt{2} - 30\sqrt{6}\right)i$$

Converting this product to polar form will be challenging (and I, for one, choose not to do all that work). But, you ask, are the answers the same number?

We can use the formula for the difference of two angles to determine the values of the cosine and sine of $\dfrac{17\pi}{12} \cdot \dfrac{17\pi}{12} = \dfrac{2\pi}{3} + \dfrac{3\pi}{4}$.

$$\cos\left(\frac{17\pi}{12}\right) = \cos\left(\frac{2\pi}{3} + \frac{3\pi}{4}\right) = \cos\left(\frac{2\pi}{3}\right)\cos\left(\frac{3\pi}{4}\right) + \sin\left(\frac{2\pi}{3}\right)\sin\left(\frac{3\pi}{4}\right)$$
$$= \frac{\sqrt{2} - \sqrt{6}}{4}$$

$$\sin\left(\frac{17\pi}{12}\right) = \sin\left(\frac{2\pi}{3} + \frac{3\pi}{4}\right) = \sin\left(\frac{2\pi}{3}\right)\cos\left(\frac{3\pi}{4}\right) + \sin\left(\frac{3\pi}{4}\right)\cos\left(\frac{2\pi}{3}\right)$$
$$= \frac{-\sqrt{6} - \sqrt{2}}{4}$$

$$120\,cis\left(\frac{17\pi}{12}\right) = 120\left(\frac{\sqrt{2} - \sqrt{6}}{4}\right) + 120i\left(\frac{-\sqrt{6} - \sqrt{2}}{4}\right)$$
$$= \left(30\sqrt{2} - 30\sqrt{6}\right) + \left(-30\sqrt{2} - 30\sqrt{6}\right)i$$

▶ Determine the product of $18\,cis\left(\dfrac{5\pi}{9}\right)$ and $3\,cis\left(\dfrac{5\pi}{18}\right)$. Write your answer in both polar and rectangular form.

▶ Polar form: $\left[18\,cis\left(\dfrac{5\pi}{9}\right)\right]\left[3\,cis\left(\dfrac{5\pi}{18}\right)\right] = 54\,cis\left(\dfrac{15\pi}{18}\right) = 54\,cis\left(\dfrac{5\pi}{6}\right)$

▶ Rectangular form: $54\,cis\left(\dfrac{5\pi}{6}\right) = 54\left(\dfrac{-\sqrt{3}}{2}\right) + 54i\left(\dfrac{1}{2}\right) = -27\sqrt{3} + 27i$

The rules for division are, as you might guess, similar to that for multiplication. The quotient is found by dividing the radii and subtracting the angles.

$$\frac{r_1\,cis(\alpha)}{r_2\,cis(\beta)} = \frac{r_1}{r_2}\,cis(\alpha - \beta)$$

▶ Find the quotient $\dfrac{z_1}{z_2}$ if $z_1 = 12\,cis\left(\dfrac{2\pi}{3}\right)$ and $z_2 = 10\,cis\left(\dfrac{3\pi}{4}\right)$.

▶ $\dfrac{z_1}{z_2} = \dfrac{12}{10}\,cis\left(\dfrac{2\pi}{3} - \dfrac{3\pi}{4}\right) = \dfrac{6}{5}\,cis\left(\dfrac{-\pi}{12}\right)$

As we saw earlier, the sine and cosine values for an angle such as $\dfrac{\pi}{12}$ can be determined exactly if we need the rectangular form. Other angles will require a decimal approximation.

▶ Find the quotient $\dfrac{z_1}{z_2}$ if $z_1 = 24\,cis\left(\dfrac{2\pi}{5}\right)$ and $z_2 = 6\,cis\left(\dfrac{\pi}{3}\right)$. Write your answer in both polar and rectangular form.

▶ Polar form: $\dfrac{z_1}{z_2} = \dfrac{24}{6}\,cis\left(\dfrac{2\pi}{5} - \dfrac{\pi}{3}\right) = 4\,cis\left(\dfrac{\pi}{15}\right)$

▶ Rectangular form: $4\,cis\left(\dfrac{\pi}{15}\right) = 4\cos\left(\dfrac{\pi}{15}\right) + 4i\sin\left(\dfrac{\pi}{15}\right) = 3.91 + 0.83i$

(Answers rounded to the nearest hundredth.)

Powers and Roots of Complex Numbers

Computing powers of complex numbers in polar form is an extension of the product rule. If $z = r\,cis(\theta)$ then z^n would be the product of $r\,cis(\theta)$ n times, yielding a final product of $r^n\,cis(n\theta)$.

EXAMPLE

▶ Compute z^5 if $z = 4\,cis\left(\dfrac{\pi}{15}\right)$. Write your answer in both polar and rectangular form.

▶ Polar form: $z^5 = 4^5\,cis\left(5\left(\dfrac{\pi}{15}\right)\right) = 1024\,cis\left(\dfrac{\pi}{3}\right)$

▶ Rectangular form:

$$1024\,cis\left(\frac{\pi}{3}\right) = 1024\left(\frac{1}{2}\right) + 1024i\left(\frac{\sqrt{3}}{2}\right) = 512 + 512i\sqrt{3}$$

Here is an example that illustrates the power of the power rule–double usage intended!

EXAMPLE

▶ Given $z = 8 - 8i\sqrt{3}$, compute z^6.

▶ You could use the formula for binomial expansion that you learned in the past, but that will be a fair amount of work. It will be easier to convert z into polar form and then apply the power rule.

▶ $z = 16\,cis\left(\dfrac{5\pi}{3}\right)$ so

$$z^6 = 16^6\,cis\left(6\left(\frac{5\pi}{3}\right)\right) = \left(2^4\right)^6 cis(10\pi) = 2^{24}(1 + 0i) = 2^{24} = 16{,}777{,}216$$

If we know the value of z^n, we can solve for a value of z by raising both sides of the equation to the reciprocal of n power. Abraham De Moivre showed that there is an easy way to find the n solution Carl Gauss proved must be there. What is particularly awesome about De Moivre's work is that it is related to finding the coordinates for the tips of the petals of the polar rose.

De Moivre's theorem says that the solution to the equation $z^n = r\,cis(\theta)$ is $z = r^{1/n}\,cis\left(\dfrac{\theta + 2\pi k}{n}\right)$, where $k = 1, 2, 3, ..., n-1$. Allow me to rewrite this with a modification, $z = r^{1/n}\,cis\left(\dfrac{\theta}{n} + \dfrac{2\pi}{n}k\right)$. The angle $\dfrac{\theta}{n}$ represents the initial value for the angle, while the fraction $\dfrac{2\pi}{n}$ tells us the rotation needed to reach the next root, and k tells us how many times we should go about the circle. But isn't the total number of roots equal to n? We already know how many times to go around the circle.

▶ Find the three cubed roots of 8.

▶ The polar form for 8 is $8\,cis(0)$ and we need to solve the equation $z^3 = 8\,cis(0)$. The first solution is $z_1 = 8^{1/3}\,cis\left(\dfrac{0}{3}\right) = 2\,cis(0) = 2$. We knew that would be an answer. I am using the subscript to keep track of all the answers. $z_2 = 2\,cis\left(0 + \dfrac{2\pi}{3}\right) = 2\,cis\left(\dfrac{2\pi}{3}\right) = -1 + i\sqrt{3}$ and $z_3 = 2\,cis\left(0 + \dfrac{2\pi}{3}(2)\right) = 2\,cis\left(\dfrac{4\pi}{3}\right) = -1 - i\sqrt{3}$.

▶ We can verify this answer because the equation $z^3 = 8$ becomes $z^3 - 8 = 0$. Factor the left-hand side to get the equation $(z-2)(z^2 + 2z + 4) = 0$, and use the quadratic formula to get the other two solutions.

I don't believe that you are likely to have seen the question prior to now.

EXAMPLE

▶ Find the fourth roots of $16i$.

▶ Write the equation $z^4 = 16\,cis\left(\dfrac{\pi}{2}\right)$. Solve this to get

$$z_1 = 16^{1/4}\,cis\left(\dfrac{\dfrac{\pi}{2}}{4}\right) = 2\,cis\left(\dfrac{\pi}{8}\right).$$ The remaining solutions are at a

rotation of $\dfrac{2\pi}{4} = \dfrac{\pi}{2}$ around the circle: $z_2 = 2\,cis\left(\dfrac{\pi}{8} + \dfrac{\pi}{2}\right) = 2\,cis\left(\dfrac{5\pi}{8}\right)$,

$z_3 = 2\,cis\left(\dfrac{9\pi}{8}\right)$, and $z_4 = 2\,cis\left(\dfrac{13\pi}{8}\right)$.

▶ Use your calculator to determine $\cos\left(\dfrac{\pi}{8}\right) = \dfrac{\sqrt{\sqrt{2}+2}}{2}$ and

$\sin\left(\dfrac{\pi}{8}\right) = \dfrac{\sqrt{2-\sqrt{2}}}{2}$. You can now determine the rectangular form for

each of the four roots of $16i$.

Let's do one more.

EXAMPLE

▶ Solve $z^5 = -16 - 16i\sqrt{3}$.

▶ The polar form for this equation is $z^5 = 32\,cis\left(\dfrac{4\pi}{3}\right)$. Each solution will

have a radius of 2. The angle for the first solution will be $\dfrac{4\pi}{15}$ and the

angle between solutions will be $\dfrac{2\pi}{5} = \dfrac{6\pi}{15}$.

▶ The five solutions are $z_1 = 2\,cis\left(\dfrac{4\pi}{15}\right)$, $z_2 = 2\,cis\left(\dfrac{2\pi}{3}\right)$, $z_3 = 2\,cis\left(\dfrac{16\pi}{15}\right)$,

$z_4 = 2\,cis\left(\dfrac{22\pi}{15}\right)$, and $z_5 = 2\,cis\left(\dfrac{28\pi}{15}\right)$.

Exponential Format

The language of mathematics has long been filled with vocabulary that has led to confusion. In this case, the notion of imaginary numbers and complex numbers strike a nerve. (We talk about real numbers, but never easy numbers or simple numbers.) I would like to finish this chapter with two computations that I think are very important in terms of recognizing that all the numbers are interrelated.

You learned about the cyclical nature of powers of i when you first studied complex numbers: $i^2 = -1$, $i^3 = -i$, $i^4 = 1$. I highly doubt that anyone gave a thought to using a complex number as an exponent.

Use your calculator to compute i^i. Are you surprised that the answer is real, not imaginary or complex? $i^i = 0.20788$. Even better, calculate $e^{-\frac{\pi}{2}}$. You get the same response.

How did this come to be?

 IRL As you are well aware, your calculator does not have a little person inside it who can compute trigonometric, logarithmic, or exponential functions with the speed of light. The computations are done with an algorithm that uses power series. (A topic you are likely to study in your second calculus course.)

The three series worth noting are for the functions e^x, $\sin(x)$, and $\cos(x)$.

$$e^x = 1 + x + \frac{x^2}{2} + \frac{x^3}{3!} + \frac{x^4}{4!} + \cdots + \frac{x^n}{n!}$$

$$\sin(x) = x - \frac{x^3}{3!} + \frac{x^5}{5!} - \frac{x^7}{7!} + \cdots + \frac{(-1)^n x^{2n+1}}{(2n+1)!}$$

$$\cos(x) = 1 - \frac{x^2}{2!} + \frac{x^4}{4!} - \frac{x^6}{6!} + \cdots + \frac{(-1)^n x^{2n}}{(2n)!}$$

The first term in each series corresponds to $n = 0$, and the notation $n!$ is read as n factorial and is equal to the product of the first n counting numbers. By definition, $0! = 1$.

It is interesting to note that if we evaluate the exponential function with the term ix, the series becomes:

$$e^{ix} = 1 + (ix) + \frac{(ix)^2}{2} + \frac{(ix)^3}{3!} + \frac{(ix)^4}{4!} + \cdots + \frac{(ix)^n}{n!}$$

$$e^{ix} = 1 + xi - \frac{x^2}{2} - \frac{x^3}{3!}i + \frac{x^4}{4!} + \frac{x^5}{5!}i - \frac{x^6}{6!} - \frac{x^7}{7!}i + \cdots + \frac{(ix)^n}{n!}$$

$$e^{ix} = \left(1 - \frac{x^2}{2!} + \frac{x^4}{4!} - \frac{x^6}{6!} + \cdots\right) + \left(x - \frac{x^3}{3!} + \frac{x^5}{5!} - \frac{x^7}{7!} + \cdots\right)i$$

$$e^{ix} = \cos(x) + i\sin(x)$$

EXAMPLE

▶ Evaluate this equation for $x = \dfrac{\pi}{2}$: $e^{\frac{i\pi}{2}} = \cos\left(\dfrac{\pi}{2}\right) + i\sin\left(\dfrac{\pi}{2}\right) = 0 + i = i$

▶ Raise both sides of the equation to the i power: $\left(e^{\frac{i\pi}{2}}\right)^i = (i)^i \Rightarrow e^{\frac{-\pi}{2}} = i^i$

We now know that i cannot only be the base of the expression a^b but also the exponent and have the result be a real number.

EXERCISES

EXERCISE 13-1

Convert each number to polar form.

1. $6 - 6i\sqrt{3}$

2. $-8\sqrt{2} - 8i\sqrt{2}$

EXERCISE 13-2

Convert each number to rectangular form.

1. $12\,cis\left(\dfrac{7\pi}{6}\right)$

2. $20\,cis\left(\dfrac{7\pi}{4}\right)$

EXERCISE 13-3

For each question, perform the indicated computations. Give answers in both polar and rectangular form.

1. $\left(18\,cis\left(\dfrac{\pi}{3}\right)\right)\left(12\,cis\left(\dfrac{5\pi}{6}\right)\right)$

2. $\left(5 + 5i\sqrt{3}\right)\left(4\sqrt{2} - 4i\sqrt{2}\right)$

3. $\left(12\,cis\left(\dfrac{\pi}{8}\right)\right)\left(24\,cis\left(\dfrac{5\pi}{24}\right)\right)$

4. $\dfrac{24\,cis\left(\dfrac{5\pi}{24}\right)}{12\,cis\left(\dfrac{\pi}{8}\right)}$

5. $\dfrac{21\sqrt{3} + 21i}{7 - 7i\sqrt{3}}$

6. $\left(6\,cis\left(\dfrac{2\pi}{3}\right)\right)^3$

7. $\left(\sqrt{2}+i\sqrt{2}\right)^8$

EXERCISE 13-4

Find all solutions to each equation.

1. $z^6 = 64\,cis\left(\dfrac{6\pi}{5}\right)$

2. $z^4 = 1+i$

EXERCISE 13-5

Rewrite each expression in $r\,cis(\theta)$ form and evaluate the expression.

1. $e^{\frac{i\pi}{3}}$

2. $e^{i\pi}$

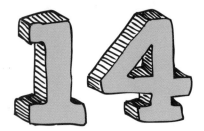

14 Introduction to Calculus

MUST KNOW

 Limits describe the behavior of a function *near* a given value of *x*, not *at* the value of *x*.

 We can now determine the rate of change of the function at a given point rather than determining the average rate of change in an interval containing the given point.

T he history of the development of calculus parallels the history of Britain versus mainland Europe in many ways. Though the mathematicians did not fire guns at each other, they did take a lot of shots at each other in the form of teasing letters, barbs, and outright antagonism. Do you know who is credited with the development of calculus? I'll bet you do. Isaac Newton is usually the person given credit. But do you know who published his findings first? Gottfried Leibnitz published his findings in 1684. It was not until 1693 that Newton, with urging from his good friend Edmund Halley (of Halley's comet), published his *Principia*. The *Principia* contained all the work Newton had been doing in physics up to that time. Why is this important? You will find that there are two sets of notation used in calculus today. One notation was developed by Newton and the other by Leibnitz. We use both notations because, depending on the application, they help us understand the concept at hand.

The Concept of a Limit

Contrary to popular myth, Newton was not hit in the head with an apple falling from a tree. He did get his idea for the basis of calculus while sitting in his den and seeing an apple fall from a tree. So, the falling apple is accurate. The rest of the legend makes for a good story.

Newton realized that the rate with which the apple was falling to the ground could be determined by computing the average velocity of the fall over shorter and shorter periods of time. His reasoning went something like this. Suppose the apple begins at a height of 25 feet above the ground. If the apple takes 1.25 seconds to fall to the ground, the average velocity of the apple in this interval is $\dfrac{25-0}{0-1.25} = -20$ feet per second (fps). The velocity is negative because the apple is falling and the distance to the ground is getting smaller. Narrow the scope of the fall to, say, the interval from when the apple was 20 feet off the ground to 10 feet off the ground. Let's assume that the time needed is 0.75 seconds. The average velocity will be $\dfrac{20-10}{0-0.35} = -28.57$ fps. He continued to reduce the length of the interval until the changes in the average velocity were insignificant. Do you see how this is a type of sequence problem that he solved?

There is a second concept that we need to discuss before we go about the business of evaluating finite limits. That concept is "close enough." Suppose you loan a friend $10, and when it is time to repay you, your friends offers you $9 to repay the loan. Do you accept it? If not that amount (and knowing that you are not going to get all of the $10 back), what do you consider "close enough" to repay the loan? Is $9.25 enough? $9.50? $9.90? $9.99? Many people will settle quickly since there is a value to your friendship that (hopefully) money cannot buy.

Let's change the scenario a little bit. Suppose the loan is not with a friend (so that does not impact your thinking) and is for $10 million. Is an offer of $9 million enough to satisfy the loan? If not that, what is close enough for you?

Here's a nonmonetary example. You are walking to a friend's house and you get a call from the friend. Your friend asks, "Where are you?" You answer, "In the neighborhood." Does that mean you are near your friend's house? How near are you?

The point to all of this is how do you determine "near," "close enough," or a term that is used to discuss limits, "in the neighborhood"? Mathematically, it is argued that $f(x)$ is in the neighborhood of a value L whenever x is near some value c. The accompanying graphic illustrates this notion.

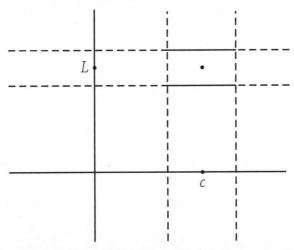

If, whenever x is close to c, $f(x)$ is close to L, we will claim that the limit as x approaches c is L and will write this as $\lim_{x \to c} f(x) = L$. Furthermore, we will define "x is close to c" as $|x - c| < \delta$. We define the entire statement as

If for all possible $\delta > 0$ with $|x - c| < \delta$, there exists a value for $\varepsilon > 0$, so $|f(x) - L| < \varepsilon \Leftrightarrow \lim_{x \to c} f(x) = L.$

The symbols δ and ε are the lowercase Greek letters delta and epsilon, and they represent very small values. The entire notation reads:

If whenever x is close to c it forces $f(x)$ to be close to L, then the limit of $f(x)$ as x approaches c is L *and* if the limit of $f(x)$ as x approaches c is L, then whenever x is close to c it forces $f(x)$ to be close to L.

BTW

This is a good place to issue a warning. We are used to working with fairly basic functions (though some might seem complex now). The theorems you will see from here on about functions must apply to all functions, not just polynomials, rational functions, and trigonometric functions. An example of a more complex function is the Dirichlet function:

$$f(x) = \begin{cases} 1, & x \text{ is rational} \\ -1, & x \text{ is irrational} \end{cases}$$

Evaluating Limits

The first step to perform when evaluating a limit as x approaches c is to substitute c for x.

EXAMPLE

▶ Evaluate $\lim_{x \to 2} 2x^3 - 3x + 2$.

▶ $\lim_{x \to 2} 2x^3 - 3x + 2 = 2(2)^3 - 3(2) + 2 = 12.$

That was easy. So is this one.

▶ Evaluate $\lim\limits_{x \to 2} \left(\dfrac{2x^3 - 3x + 2}{x^2 + 4} \right)$.

▶ $\lim\limits_{x \to 2} \left(\dfrac{2x^3 - 3x + 2}{x^2 + 4} \right) = \dfrac{2(2)^3 - 3(2) + 2}{2^2 + 4} = \dfrac{12}{8} = \dfrac{3}{2}$.

Let's look at something a bit more challenging.

▶ Evaluate $\lim\limits_{x \to 2} \left(\dfrac{2x^3 - 3x - 10}{x^2 - 4} \right)$.

▶ Granted, this doesn't look much different than the last problem. That is, it doesn't look much different until we substitute $x = 2$ into the expression.

$$\lim\limits_{x \to 2} \left(\dfrac{2x^3 - 3x - 10}{x^2 - 4} \right) = \dfrac{2(2)^3 - 3(2) - 10}{2^2 - 4} = \dfrac{0}{0}.$$

▶ We have an interesting situation here. We know that $\dfrac{0}{a} = 0$, provided that a is not equal to zero and that $\dfrac{a}{0}$ is undefined. We call the expression $\dfrac{0}{0}$ **indeterminate** because we cannot determine the value of the expression in its current form. We can apply the Factor theorem to both the numerator and denominator since we know that $x - 2$ is a factor of each.

$$\lim\limits_{x \to 2} \left(\dfrac{2x^3 - 3x - 10}{x^2 - 4} \right) = \lim\limits_{x \to 2} \left(\dfrac{(x - 2)(2x^2 + 4x + 5)}{(x - 2)(x + 2)} \right)$$

$$= \lim\limits_{x \to 2} \left(\dfrac{2x^2 + 4x + 5}{x + 2} \right) = \dfrac{2(2)^2 + 3(2) + 5}{2 + 2} = \dfrac{19}{4}$$

This next example is a classic.

EXAMPLE

▶ Evaluate $\lim\limits_{x \to 4}\left(\dfrac{x-4}{\sqrt{x}-2}\right)$.

▶ We get the indeterminate form $\dfrac{0}{0}$ when we substitute 4 into the expression. The Factor theorem tells us that $\sqrt{x}-2$ is a factor of both the numerator and denominator. However, have you ever factored $x-4$? Well, you are going to now. Did you know that you can treat $\sqrt{x}-2$ as the difference of squares?

▶ $\lim\limits_{x \to 4}\left(\dfrac{x-4}{\sqrt{x}-2}\right) = \lim\limits_{x \to 4}\left(\dfrac{\left(\sqrt{x}-2\right)\left(\sqrt{x}+2\right)}{\sqrt{x}-2}\right) = \lim\limits_{x \to 4}\left(\sqrt{x}+2\right) = 4$

As strange as it seems to factor a linear expression, our objective is to remove the common factor from the numerator and denominator. Yes, it is true that you could have rationalized the denominator by multiplying both the numerator and denominator by $\sqrt{x}+2$. However, this was a chance to show an interesting application of factoring that also reduces the number of steps needed to solve the problem.

Be careful when you see the indeterminate form and simplify the expression.

EXAMPLE

▶ Evaluate $\lim\limits_{x \to 4}\left(\dfrac{x-4}{x^2-16}\right)$.

▶ We get the indeterminate form when we substitute 4 into the expression. After factoring the denominator and reducing the fraction, the problem now reads $\lim\limits_{x \to 4}\left(\dfrac{1}{x-4}\right) = \dfrac{1}{0}$. We can't divide 1 by 0, so what does this tell us?

▶ Remember, we are asking how the function behaves in the neighborhood of $x=4$. To do this, we need to look at both sides of the neighborhood. That is, we need to look at the behavior of the function for values of x that are less than 4 and for values of x that are greater than 4. We saw this material back in Chap. 4 when we studied rational functions. As x approaches 4 from the left side of the number line, the graph goes to negative infinity, while the behavior of the function for values of x greater than 4 is that the graph goes to positive infinity. We call these left- and right-hand limits.

▶ Left-hand limit: $\lim\limits_{x \to 4^-} \left(\dfrac{1}{x - 4} \right) = -\infty.$

   Right-hand limit: $\lim\limits_{x \to 4^+} \left(\dfrac{1}{x - 4} \right) = \infty.$

▶ Because these limits are not the same, we state that $\lim\limits_{x \to 4} \left(\dfrac{x - 4}{x^2 - 16} \right)$ is DNE (does not exist).

What do we write as an answer if both the left- and right-hand limits are the same? If the limits are finite values, then this number is our limiting value. If the value of the limit is that it is infinitely large, then we state that the limit is infinitely large. The limit is supposed to tell us behavior, not a value, so if we can reveal that both branches of a curve near an asymptote are tracking in the same direction, we do so by stating the limit is infinitely large.

EXAMPLE

▶ Evaluate $\lim\limits_{x \to 4} \left(\dfrac{x^2 - 16}{(x - 4)^3} \right)$

$$\lim\limits_{x \to 4} \left(\dfrac{x^2 - 16}{(x - 4)^3} \right) = \lim\limits_{x \to 4} \left(\dfrac{(x - 4)(x + 4)}{(x - 4)^3} \right) = \lim\limits_{x \to 4} \left(\dfrac{x + 4}{(x - 4)^2} \right) = \dfrac{8}{0}.$$

▶ Compute the one-sided limits. Left: $\lim\limits_{x \to 4^-} \left(\dfrac{x + 4}{(x - 4)^2} \right) = \infty$; right:
   $\lim\limits_{x \to 4^+} \left(\dfrac{x + 4}{(x - 4)^2} \right) = \infty$

▶ Therefore, $\lim\limits_{x \to 4} \left(\dfrac{x^2 - 16}{(x - 4)^3} \right) = \infty.$

We should look at one piecewise function while talking about one-sided limits.

EXAMPLE

▶ Evaluate $\lim\limits_{x \to -2} f(x)$ if $f(x) = \begin{cases} x+3, & x <= -2 \\ 5 & x = -2 \\ x^2 - 3 & x > -2 \end{cases}$

▶ Clearly we need to use one-sided limits for this problem since the point at which the limit is being evaluated is also the point at which the domain is split.

▶ Left: $\lim\limits_{x \to -2^-} f(x) = \lim\limits_{x \to -2^-} x + 3 = 1$

Right: $\lim\limits_{x \to -2^+} f(x) = \lim\limits_{x \to -2^+} x^2 - 3 = 1$

▶ The left-hand limit is equal to the right-hand limit, so we have $\lim\limits_{x \to -2} f(x) = 1$.

This is a good time to note that the limiting value, 1, is not the same as the functional value, 5, when $x = -2$.

We discussed limits to infinity in Chap. 4 when we discussed the end behavior of rational functions. If the function is a polynomial function, the limiting value will be either positive or negative infinity and is dependent on the highest degree of the polynomial and the sign of the leading coefficient.

EXAMPLE

▶ Evaluate $\lim\limits_{x \to -\infty} (-2x^3 + 8x)$.

▶ A negative number cubed will be negative and when multiplied by the negative coefficient will yield a positive result.
Therefore, $\lim\limits_{x \to -\infty} (-2x^3 + 8x) = \infty$.

Here is an example of limits to infinity with a rational function.

▶ Evaluate $\lim\limits_{x \to \infty} \dfrac{4x^3 - 7x^2 + 12x - 9}{5x^3 + 12x - 200}$.

▶ The degree of the numerator is equal to the degree of the denominator.
Dividing both by x^3 gives

$$\lim_{x \to \infty} \frac{4 - \dfrac{7}{x} + \dfrac{12}{x^2} - \dfrac{9}{x^3}}{5 + \dfrac{12}{x^2} - \dfrac{200}{x^3}} = \frac{4}{5} \text{ because all fractional terms will go to}$$

zero as the value of x gets infinitely large.

There is a statement in calculus that is called the Pinching theorem (or sometimes the Squeeze theorem). The theorem states that if there is an interval for which three functions, $f(x)$, $g(x)$, and $k(x)$, satisfy the criteria $f(x) \le g(x) \le k(x)$ and $\lim\limits_{x \to 0} f(x) = \lim\limits_{x \to 0} k(x) = c$, then $\lim\limits_{x \to 0} g(x) = c$.

This theorem is used to prove a limit statement that is crucial to most of the limits involving trigonometric functions, $\lim\limits_{h \to 0} \dfrac{\sin(h)}{h} = 1$.

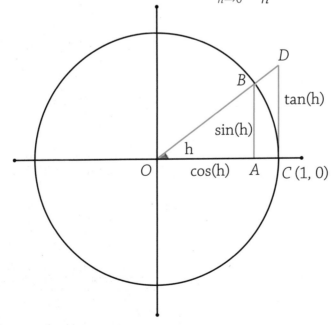

- The diagram of this unit circle shows that

$$\text{area } \triangle AOB \leq \text{area sector } OCB \leq \text{area } \triangle COD.$$

- Substituting the area formulas for the figures, the inequality becomes

$$\frac{1}{2}\sin(h)\cos(h) \leq \frac{1}{2}(1)^2 h \leq \frac{1}{2}\tan(h).$$

- Multiply all three expressions by 2 and use the quotient identity for the tangent statement.

$$\sin(h)\cos(h) \leq h \leq \frac{\sin(h)}{\cos(h)}$$

- Divide each expression by $sin(h)$.

$$\cos(h) \leq \frac{h}{\sin(h)} \leq \frac{1}{\cos(h)}$$

- Take the reciprocal of all three expressions, knowing that we also need to change the orientation of the inequalities.

$$\frac{1}{\cos(h)} \geq \frac{\sin(h)}{h} \geq \cos(h)$$

- Use the fact that $\lim\limits_{h \to 0} \dfrac{1}{\cos(h)} = \lim\limits_{h \to 0} \cos(h) = 1$

  to prove $\lim\limits_{h \to 0} \dfrac{\sin(h)}{h} = 1$.

BTW

Another important limit statement (which we will not prove at this time)

is $\lim\limits_{n \to \infty} \left(1 + \frac{x}{n}\right)^n = e^x$.

Property of Limits

The following are properties of limits that apply to both one-sided and two-sided limits.

If $L_1 = \lim\limits_{x \to c} f(x)$ and $L_2 = \lim\limits_{x \to c} g(x)$ then

- $\lim\limits_{x \to c} [f(x) + g(x)] = \lim\limits_{x \to c} f(x) + \lim\limits_{x \to c} g(x) = L_1 + L_2$

- $\lim\limits_{x \to c} [f(x) - g(x)] = \lim\limits_{x \to c} f(x) - \lim\limits_{x \to c} g(x) = L_1 - L_2$

- $\lim\limits_{x \to c} [f(x) \times g(x)] = \lim\limits_{x \to c} f(x) \times \lim\limits_{x \to c} g(x) = L_1 \times L_2$

- $\lim\limits_{x \to c} \left[\dfrac{f(x)}{g(x)}\right] = \dfrac{\lim\limits_{x \to c} f(x)}{\lim\limits_{x \to c} g(x)} = \dfrac{L_1}{L_2}$

- $\lim\limits_{x \to c} kf(x) = k \lim\limits_{x \to c} f(x) = kL_1$

- $\lim\limits_{x \to c} \sqrt[n]{f(x)} = \sqrt[n]{\lim\limits_{x \to c} f(x)} = \sqrt[n]{L_1}$

- $\lim\limits_{x \to c} f(g(x)) = f(\lim\limits_{x \to c} g(x)) = f(L_2)$

EXAMPLE

▶ Use $\lim\limits_{x \to 0} \dfrac{\sin(x)}{x} = 1$, what is $\lim\limits_{x \to 0} \dfrac{\sin(2x)}{x}$?

▶ $\lim\limits_{x \to 0} \dfrac{\sin(2x)}{x} = \lim\limits_{x \to 0} \dfrac{2\sin(x)\cos(x)}{x} = \lim\limits_{x \to 0} \dfrac{2\sin(x)}{x} \lim\limits_{x \to 0} \cos(x)$

$$= 2\lim\limits_{x \to 0} \dfrac{\sin(x)}{x} \lim\limits_{x \to 0} \cos(x) = 2(1)(1) = 2$$

Here is another example.

EXAMPLE

▶ Evaluate $\lim\limits_{x \to 0} \dfrac{\tan(x)}{x}$.

▶ $\lim\limits_{x \to 0} \dfrac{\tan(x)}{x} = \lim\limits_{x \to 0} \dfrac{\frac{\sin(x)}{\cos(x)}}{x} = \lim\limits_{x \to 0} \dfrac{\sin(x)}{x\cos(x)}$

$$= \lim\limits_{x \to 0} \dfrac{\sin(x)}{x} \lim\limits_{x \to 0} \dfrac{1}{\cos(x)} = (1)(1) = 1.$$

Here is one more example.

EXAMPLE

▶ Evaluate $\lim\limits_{h \to 0} \dfrac{\sin\left(\dfrac{\pi}{3} + h\right) - \sin\left(\dfrac{\pi}{3}\right)}{h}$.

▶ $\lim\limits_{h \to 0} \dfrac{\sin\left(\dfrac{\pi}{3} + h\right) - \sin\left(\dfrac{\pi}{3}\right)}{h} = \lim\limits_{h \to 0} \dfrac{\sin\left(\dfrac{\pi}{3}\right)\cos(h) + \sin(h)\cos\left(\dfrac{\pi}{3}\right) - \sin\left(\dfrac{\pi}{3}\right)}{h}$

$\qquad = \lim\limits_{h \to 0} \dfrac{\sin\left(\dfrac{\pi}{3}\right)[\cos(h) - 1]}{h} + \lim\limits_{h \to 0} \dfrac{\cos\left(\dfrac{\pi}{3}\right)\sin(h)}{h}$

$\qquad = \sin\left(\dfrac{\pi}{3}\right)\lim\limits_{h \to 0} \dfrac{\cos(h) - 1}{h} + \cos\left(\dfrac{\pi}{3}\right)\lim\limits_{h \to 0} \dfrac{\sin(h)}{h}$

▶ We were told in the last two examples that $\lim\limits_{x \to 0} \dfrac{\sin(x)}{x} = 1$.

Changing the x to h does not change the value of the limit, so

$\cos\left(\dfrac{\pi}{3}\right)\lim\limits_{h \to 0} \dfrac{\sin(h)}{h} = \left(\dfrac{1}{2}\right)(1) = \dfrac{1}{2}$. We also have $\lim\limits_{h \to 0} \dfrac{\cos(h) - 1}{h} = \dfrac{0}{0}$,

an indeterminate form.

▶ We cannot factor this expression, so we will have to apply some trigonometric identity. Multiply the numerator and denominator by $\cos(h) + 1$,

$\lim\limits_{h \to 0} \left(\dfrac{\cos(h) - 1}{h}\right)\left(\dfrac{\cos(h) + 1}{\cos(h) + 1}\right) = \lim\limits_{h \to 0} \dfrac{\cos^2(h) - 1}{h\cos(h) + 1} = -\lim\limits_{h \to 0} \dfrac{1 - \cos^2(h)}{h(\cos(h) + 1)}$.

▶ Using the Pythagorean identity, we get

$\lim\limits_{h \to 0} \left(\dfrac{\cos(h) - 1}{h}\right) = -\lim\limits_{h \to 0} \dfrac{\sin^2(h)}{h(\cos(h) + 1)} = -\lim\limits_{h \to 0} \left[\dfrac{\sin(h)}{h} \dfrac{\sin(h)}{\cos(h) + 1}\right]$

$\qquad = -\lim\limits_{h \to 0} \left[\dfrac{\sin(h)}{h}\right] \times \lim\limits_{h \to 0} \left[\dfrac{\sin(h)}{\cos(h) + 1}\right] = -(1)\left(\dfrac{0}{2}\right) = 0.$

▶ Therefore, $\lim\limits_{h \to 0} \dfrac{\sin\left(\dfrac{\pi}{3} + h\right) - \sin\left(\dfrac{\pi}{3}\right)}{h} = \dfrac{1}{2}$.

Continuity

What does it mean for a function to be continuous? A layperson's definition might be this: a function that can be drawn so that, once the pencil is put to paper, the pencil never has to leave the paper to finish the graph.

Mathematicians have a more precise definition. First, for a function $f(x)$ to be **continuous** at a point $x = c$, it must be the case that $\lim_{x \to c} f(x) = f(c)$.

▶ Is the function $f(x) = \begin{cases} x + 3, & x <= -2 \\ 5 & x = -2 \\ x^2 - 3 & x > -2 \end{cases}$ continuous at $x = -2$?

▶ The answer is no. We showed earlier that $\lim_{x \to -2} f(x) = 1$. However, $f(-2) = 5$, so the function is not continuous.

This type of discontinuity is called a **removable discontinuity**. It is removable because we could define the function to have a value of 1 at $x = -2$ rather than 5. Recall that in Chap. 4 we also talked about **jump discontinuities** and **infinite discontinuities**.

The second statement mathematicians have about continuity is this: a function is said to be continuous if it is continuous at all values of x.

▶ For what value of c will the function $f(x) = \begin{cases} \dfrac{x^2 - 25}{x - 5}, & x \neq 5 \\ c & x = 5 \end{cases}$

be continuous?

▶ Knowing that $f(5) = \lim_{x \to 5} f(x) = \lim_{x \to 5} \dfrac{x^2 - 25}{x - 5} = \lim_{x \to 5} \dfrac{(x - 5)(x + 5)}{x - 5} =$
$\lim_{x \to 5} x + 5 = 10$, c must be equal to 10.

Polynomial functions are continuous.

Instantaneous Rates of Change

A tangent line to a circle is a line that intersects the circle at one point. We can also define a line that intersects a curve at a single point (in the neighborhood of the point) without crossing through the curve to be a tangent line. The process we use to determine the slope of this tangent line is the same process Newton used to determine the velocity of the falling apple all those years ago.

EXAMPLE

▶ Let's determine the slope of the line tangent to the graph of the function $y = f(x) = x^2$ at the point (3, 9). We need a second point to determine the slope of a line. We will pick a point that is near 3, $(3 + \Delta x, (3 + \Delta x)^2)$. The slope of the line joining these two points is

$$\frac{(3 + \Delta x)^2 - 9}{3 + \Delta x - 3}.$$

▶ Simplify this expression.

$$\frac{9 + 6\Delta x + (\Delta x)^2 - 9}{3 + \Delta x - 3} = \frac{6\Delta x + (\Delta x)^2}{\Delta x} = \frac{\Delta x(6 + \Delta x)}{\Delta x} = 6 + \Delta x$$

▶ If we take the limit of this expression as $\Delta x \to 0$, the slope of the tangent line is 6.

In general, we can find the slope of the tangent line at a point $x = c$ by applying the slope relation:

$$\lim_{\Delta x \to 0} \frac{f(c + \Delta x) - f(c)}{\Delta x}$$

EXAMPLE

▶ Determine the slope of the tangent line to the graph of $f(x) = 5x^2 + 6x - 3$ at the point when $x = -2$.

▶ Apply the limit to the slope statement.

$$\lim_{\Delta x \to 0} \frac{5(-2 + \Delta x)^2 + 6(-2 + \Delta x) - 3 - 5}{\Delta x}$$

▶ Simplify the expression.

$$\lim_{\Delta x \to 0} \frac{20 - 10\Delta x + 5(\Delta x)^2 - 12 + 6(\Delta x) - 8}{\Delta x}$$

$$\lim_{\Delta x \to 0} \frac{-4\Delta x + 5(\Delta x)^2}{\Delta x}$$

$$\lim_{\Delta x \to 0} \frac{\Delta x(-4 + 5\Delta x)}{\Delta x} = \lim_{\Delta x \to 0} -4 + 5\Delta x = -4$$

Let's apply the formula for determining the slope of the tangent line to a rational function.

▶ Determine the slope of the tangent line to the graph of $f(x) = \dfrac{x + 1}{2x - 1}$ at the point when $x = 4$.

▶ Warning: The algebra is going to get a bit involved. Get a pencil and a piece of paper and follow along as we do this.

▶ Apply the limit to the slope statement.

$$\lim_{\Delta x \to 0} \frac{\dfrac{(4 + \Delta x) + 1}{2(4 + \Delta x) - 1} - \dfrac{-5}{7}}{\Delta x}$$

$$\lim_{\Delta x \to 0} \frac{\dfrac{\Delta x + 5}{2\Delta x - 7} + \dfrac{5}{7}}{\Delta x}$$

▶ This is a complex fraction. Multiply through the numerator and denominator by the common denominator $7(2\Delta x - 7)$.

$$\lim_{\Delta x \to 0} \frac{\left(\dfrac{\Delta x + 5}{2\Delta x - 7} + \dfrac{5}{7}\right)}{\Delta x} \left(\frac{7(2\Delta x - 7)}{7(2\Delta x - 7)}\right) = \lim_{\Delta x \to 0} \frac{7(\Delta x + 5) + 5(2\Delta x - 7)}{7\Delta x(2\Delta x - 7)}$$

▸ Clean up the fraction.

$$\lim_{\Delta x \to 0} \frac{7\Delta x + 35 - 10\Delta x - 35}{7\Delta x(2\Delta x - 7)} = \lim_{\Delta x \to 0} \frac{-3\Delta x}{7\Delta x(2\Delta x - 7)}$$

$$= \lim_{\Delta x \to 0} \frac{-3}{7(2\Delta x - 7)} = \frac{3}{49}$$

Let's do one more example. This time we examine an irrational function.

▸ Determine the slope of the tangent line to the graph of $f(x) = \sqrt{2x + 7}$ at the point when $x = 9$.

▸ Apply the limit to the slope statement.

$$\lim_{\Delta x \to 0} \frac{\sqrt{2(9 + \Delta x) + 7} - 5}{\Delta x}$$

▸ Multiply the numerator and denominator of the fraction by the conjugate of the numerator.

$$\lim_{\Delta x \to 0} \left(\frac{\sqrt{2(9 + \Delta x) + 7} - 5}{\Delta x} \right)\left(\frac{\sqrt{2(9 + \Delta x) + 7} + 5}{\sqrt{2(9 + \Delta x) + 7} + 5} \right)$$

▸ Simplify the numerator.

$$\lim_{\Delta x \to 0} \frac{2(9 + \Delta x) + 7 - 25}{\Delta x\left(\sqrt{2(9 + \Delta x) + 7} + 5\right)} = \lim_{\Delta x \to 0} \frac{18 + 2\Delta x + 7 - 25}{\Delta x\left(\sqrt{2(9 + \Delta x) + 7} + 5\right)}$$

$$= \lim_{\Delta x \to 0} \frac{2\Delta x}{\Delta x\left(\sqrt{2(9 + \Delta x) + 7} + 5\right)} = \lim_{\Delta x \to 0} \frac{2}{\left(\sqrt{2(9 + \Delta x) + 7} + 5\right)}$$

$$= \frac{2}{10} = \frac{1}{5}$$

You can see how the algebra skills you've learned in the past few years really come into play in determining the values of these limits.

The last example we gave in the section on evaluating limits asked us to

evaluate $\lim\limits_{h \to 0} \dfrac{\sin\left(\dfrac{\pi}{3} + h\right) - \sin\left(\dfrac{\pi}{3}\right)}{h}$. Do you see that if we replace h with Δx,

the problem asks us to find the slope of the line tangent to the function

$f(x) = \sin(x)$ at $x = \dfrac{\pi}{3}$? The process for finding the slope of the tangent line

works for all functions.

Differentiation Rules

While it is good to know that we can work through the limit statement
to determine the slope of a tangent line to a curve at a given value of x, it
would be a lot easier if we could apply a formula to determine the slope of
the tangent line at any value of x. Before we do that, we need to change our
vocabulary and discuss some notation. The expression used to determine
the slope of the tangent line, or the instantaneous velocity, is called the
derivative. Given that, we define the derivative as being

$$\lim\limits_{\Delta x \to 0} \dfrac{f(x + \Delta x) - f(x)}{\Delta x}.$$

It has come to be that Δx has been replaced with h, so the formula is also
written as:

$$\lim\limits_{h \to 0} \dfrac{f(x + h) - f(x)}{h}$$

This new notation has one less symbol that we need to write.

Newton referred to this notion of the instantaneous rate of change
as a fluxion. His notation for the fluxion related to the function $f(x)$
was to affix a dot on top of the notation, $\dot{f}$. This was an issue for the
printers, and the prime notation was substituted. The notation that is
used is $f'(x)$. Leibnitz saw it differently. The notation for slope had been

$\dfrac{\text{Change in } y}{\text{Change in } x} = \dfrac{\Delta y}{\Delta x}$. He wanted to stress that this was still the case by showing that the changes were infinitesimally small. He changed the use of Δ to d, creating the notation for the derivative as $\dfrac{dy}{dx}$. We use both notations still.

- Determine the derivative for the function $f(x) = x^n$, where n is a positive integer. Apply the definition of the derivative.

$$\frac{dy}{dx} = f'(x) = \lim_{h \to 0} \frac{f(x+h) - f(x)}{h} = \lim_{h \to 0} \frac{(x+h)^n - x^n}{h}$$

- The expansion for $(x + h)^n$ is

$$(x+h)^n = x^n + nx^{n-1}h + \frac{n(n-1)}{2}x^{n-2}h^2 + \frac{n(n-1)(n-2)}{3!}x^{n-3}h^3$$
$$+ \cdots + h^n.$$

- When x^n is subtracted from this expansion, all the remaining terms have a factor of h. Factor h from these terms. We now have

$$\frac{dy}{dx} = f'(x) = \lim_{h \to 0} \frac{f(x+h) - f(x)}{h}$$
$$= \lim_{h \to 0} \frac{h\left(nx^{n-1} + \frac{n(n-1)}{2}x^{n-2}h + \cdots + h^{n-1}\right)}{h}.$$

- Reduce the fraction.

$$\frac{dy}{dx} = f'(x) = \lim_{h \to 0}\left(nx^{n-1} + \frac{n(n-1)}{2}x^{n-2}h + \cdots + h^{n-1}\right)$$

- As h approaches zero, all terms but the first drop out of the problem.

- Therefore, we have if $y = f(x) = x^n$ then $\dfrac{dy}{dx} = f'(x) = nx^{n-1}$.

EXAMPLE

▶ Determine the derivative for $f(x) = x^9$.

▶ $f'(x) = 9x^8$

We can apply the rules for limits to determine the derivative for polynomial functions. We have

- $\lim\limits_{x \to c} [f(x) + g(x)] = \lim\limits_{x \to c} f(x) + \lim\limits_{x \to c} g(x) = L_1 + L_2$
- $\lim\limits_{x \to c} [f(x) - g(x)] = \lim\limits_{x \to c} f(x) - \lim\limits_{x \to c} g(x) = L_1 - L_2$
- $\lim\limits_{x \to c} kf(x) = k \lim\limits_{x \to c} f(x) = kL_1$

Therefore, we can differentiate a polynomial simply by differentiating each of the terms of the polynomial.

EXAMPLE

▶ Find the derivative of the function $y = 5x^7 + 3x^4 - 8x - 7$.

$$\frac{dy}{dx} = 35x^6 + 12x^3 - 8$$

The derivative of x^7 is $7x^6$, so the derivative of 5 times that amount is 5 times the derivative. In addition, the derivative of the constant -7 is zero, since the numerator in the fraction in the definition of the derivative will be zero, making the limiting value zero.

EXAMPLE

▶ Determine the equation for the line tangent to $f(x) = x^4 - 18x^2 - 13$ at $x = 4$.

▶ The derivative for the function is $f'(x) = 4x^3 - 36x$. The slope of the tangent line is $f'(4) = 4(4)^3 - 36(4) = 256 - 144 = 112$.

▶ When $x = 4$, $y = -45$, so the equation of the line tangent to $f(x)$ at $x = 4$ is $y + 45 = 112(x - 3)$.

Here is an important variation of this same problem.

▶ Determine the values of x for which the tangent to the function $f(x) = x^4 - 18x^2 - 13$ is a horizontal line.

▶ Given $f'(x) = 4x^3 - 36x = 4x(x^2 - 9)$, we see that $f'(x) = 0$ when $x = 0, -3, 3$.

Why is this an important variation? Examine the graph of the function.

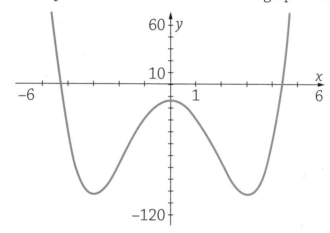

The turning points for the graph of the function occur at these points. Observe that for $x < -3$ and for $0 < x < 3$, the slopes of the tangent lines are negative, while the slopes of the tangent lines are positive when $-3 < x < 0$ and $x > 3$.

When the slopes go from negative to positive, the point where the graph turns is a relative minimum, and when the slopes go from positive to negative, the turning point is a relative maximum. This is an extremely important application of the derivative. Be sure you understand what this statement says. If there is a turning point, the value of the derivative at that point will be zero. It does *not* say that if the value of the derivative is zero, there will be a turning point at that location.

The function $g(x) = x^4 - x^3$ has a derivative $g'(x) = 4x^3 - 3x^2 = x^2(4x - 3)$, telling us that $g'(x) = 0$ when $x = 0, \dfrac{4}{3}$. The graph of $y = g(x)$ shows us that at $x = 0$, the graph does something interesting.

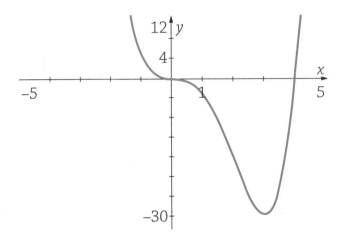

EXAMPLE

▶ Determine the coordinates of the graph of the function $f(x) = 4x^3 - 5x^2 - 8x + 3$.

▶ Find the derivative: $f'(x) = 12x^2 - 10x - 8$

▶ Solve: $f'(x) = 0$: $2(2x + 1)(3x - 4) = 0 \Rightarrow x = \dfrac{-1}{2}, \dfrac{4}{3}$

▶ Check that the signs of the derivative change about these critical points: $f'(-1) > 0; f'(0) < 0; f'(2) > 0$.

▶ Get the y-coordinates: $f\left(\dfrac{-1}{2}\right) = \dfrac{21}{4}$, $f\left(\dfrac{4}{3}\right) = \dfrac{-191}{27}$. The function has a relative maximum at $\left(\dfrac{-1}{2}, \dfrac{21}{4}\right)$ and a relative minimum at $\left(\dfrac{4}{3}, \dfrac{-191}{27}\right)$.

EXAMPLE

▶ Determine the derivative of the function $f(x) = \sin(x)$.

▶ We will need to use the definition of the derivative to determine this answer.

$$f'(x) = \lim_{h \to 0} \frac{\sin(x + h) - \sin(x)}{h}$$

$$= \lim_{h \to 0} \frac{\sin(x)\cos(h) + \sin(h)\cos(x) - \sin(x)}{h}$$

$$= \lim_{h \to 0} \frac{\sin(x)(\cos(h) - 1) + \sin(h)\cos(x)}{h}$$

$$= \lim_{h \to 0} \frac{\sin(x)(\cos(h) - 1)}{h} + \lim_{h \to 0} \frac{\sin(h)\cos(x)}{h}$$

▶ We showed earlier in this chapter that $\lim_{h \to 0} \frac{\cos(h) - 1}{h} = 0$, so we now have

$$f'(x) = 0 + \cos(x)\lim_{h \to 0} \frac{\sin(h)}{h} = \cos(x)(1) = \cos(x).$$

There are some differentiation formulas whose derivations you will see in your first calculus class but whose applications are worth showing now. These formulas are

- **Product rule**

    Newton: If $f(x) = g(x)k(x)$, then $f'(x) = g'(x)k(x) + g(x)k'(x)$.

    Leibnitz: If $y = uv$, then $\dfrac{dy}{dx} = v\dfrac{du}{dx} + u\dfrac{dv}{dx}$.

    The derivative of the product of two functions is equal to the derivative of the first function times the second function plus the first function times the derivative of the second function.

▶ Find the derivative of the function $f(x) = x\sin(x)$.

▶ Let $g(x) = x$ and $k(x) = \sin(x)$. We then

have $f'(x) = [1]\sin(x) + x[\cos(x)] = \sin(x) + x\cos(x)$.

▶ I use the brackets to highlight the derivative being taken. This is not a standard practice.

- ### Quotient rule

  Newton: If $f(x) = \dfrac{g(x)}{k(x)}$, then $f'(x) = \dfrac{g'(x)k(x) - g(x)k'(x)}{(k(x))^2}$.

  Leibnitz: If $y = \dfrac{u}{v}$ then $\dfrac{dy}{dx} = \dfrac{v\dfrac{du}{dx} - u\dfrac{dv}{dx}}{v^2}$.

  The derivative of the quotient of two functions is the derivative of the first function times the second function minus the first function times the derivative of the second function all over the square of the second function.

▸ If $y = \dfrac{x^2 + 3}{2x^3 + x}$,

▸ then $\dfrac{dy}{dx} = \dfrac{(2x^3 + x)[2x] - (x^2 + 3)[6x^2 + 1]}{(2x^3 + x)^2}$

$\qquad = \dfrac{-2x^4 - 17x^2 - 3}{(2x^3 + x)^2}$.

- ### Power rule

  If n is a real number and $y = f(x) = x^n$, then $\dfrac{dy}{dx} = f'(x) = nx^{n-1}$.

  The Power rule is really a pretty big deal because it tells us that

  $$\frac{d}{dx}(\sqrt{x}) = \frac{1}{2}x^{\frac{-1}{2}} = \frac{1}{2\sqrt{x}} \text{ as well as that}$$

  $$\frac{d}{dx}\left(\frac{1}{x^4}\right) = \frac{d}{dx}(x^{-4}) = -4x^{-5} = \frac{-4}{x^5}.$$

  There is one more rule that we need to examine. This rule is extremely important because it tells us how to take the derivative of the composition of functions.

■ **Chain rule**

Newton: If $f(x) = g(k(x))$ then $f'(x) = g'(k(x))k'(x)$.

Leibnitz: If $y = u(v)$ then $\dfrac{dy}{dx} = \dfrac{du}{dv}\dfrac{dv}{dx}$.

The derivative of a composed function is equal to the product of the derivative of the outer function evaluated at the inner function and the derivative of the inner function.

EXAMPLE

▶ Find the derivative of $f(x) = (4x^2 + 3x - 2)^5$.

▶ The outer function is $g(x) = x^5$ and the inner function is $k(x) = 4x^2 + 3x - 2$. The derivative is $f'(x) = 5(4x^2 + 3x - 2)^4[8x + 3]$.

This next example requires the use of a trigonometric identity as well as the Chain rule.

EXAMPLE

▶ Find the derivative of $y = \cos(x)$.

▶ Yes, we could go back to the definition of the derivative to determine the answer to this problem. However, we can take advantage of the cofunction identity $\cos(x) = \sin\left(\dfrac{\pi}{2} - x\right)$.

▶ The derivative of $\sin\left(\dfrac{\pi}{2} - x\right)$ is $\cos\left(\dfrac{\pi}{2} - x\right)[-1] = -\sin(x)$.

$$\dfrac{d}{dx}(\cos(x)) = -\sin(x)$$

Here is an example that puts all of the rules to use.

▶ Find the derivative of the function $p(x) = \dfrac{x \sin^2(4x)}{\sqrt{x^4 - 2}}$.

▶ Let's talk about this function before we try to differentiate it. Clearly, there is a quotient and there is a product.

▶ Let's slow the process down a little bit and just take the derivative of the numerator and then slow the process even further by examining the derivative of the expression $\sin^2(4x)$.

▶ This is the composition of three functions: x^2, $\sin(x)$, and $4x$. The derivative of the outer function is $2x$, which when evaluated with the inner function is $2\sin(4x)$. The derivative of the middle function is $\cos(x)$, which when evaluated with the inner function is $\cos(4x)$. The derivative of the innermost function is 4. Therefore, the derivative of $\sin^2(4x)$ is $8\sin(4x)\cos(4x)$.

▶ The derivative of the numerator is

$$\frac{d}{dx}(x\sin^2(4x)) = [1]\sin^2(4x) + x[8\sin(4x)\cos(4x)].$$

▶ The derivative of the denominator is

$$\frac{d}{dx}\left((x^4 - 2)^{\frac{1}{2}} \right) = \frac{1}{2}(x^4 - 2)^{\frac{-1}{2}}(4x^3) = \frac{2x^3}{\sqrt{x^4 - 2}}.$$

▶ We finally have

$$p'(x) = \frac{\left[\sin^2(4x) + 8x\sin(4x)\cos(4x) \right]\sqrt{x^4 - 2} - (x\sin^2(4x))\dfrac{2x^3}{\sqrt{x^4 - 2}}}{x^4 - 2}.$$

▶ Since the purpose of this problem was to illustrate how all the differentiation rules can be applied in one problem, it is not surprising that the result is a very complicated algebraic expression, the likes of which you will never have to simplify.

EXERCISES

EXERCISE 14-1

Evaluate each of the given limits.

1. $\displaystyle\lim_{x \to 5} \frac{x^2 - 10x + 25}{x^2 - 25}$

2. $\displaystyle\lim_{x \to 9} \frac{\sqrt{x} - 3}{x - 9}$

3. $\displaystyle\lim_{x \to \infty} \frac{3x^2 - 10x + 25}{4x^2 - 25}$

4. $\displaystyle\lim_{x \to -2} f(x)$ given $f(x) = \begin{cases} x^2 - 1, & x < -2 \\ 5, & x = -2 \\ 4x + 11, & x > -2 \end{cases}$

5. $\displaystyle\lim_{x \to 3} g(x)$ given $g(x) = \begin{cases} x^2 - 1, & x < 3 \\ 9, & x = 3 \\ 13 - x, & x > 3 \end{cases}$

6. $\displaystyle\lim_{x \to 0} \frac{\tan(2x)}{x}$

7. $\displaystyle\lim_{\Delta x \to 0} \frac{\cos\left(\dfrac{\pi}{6} + \Delta x\right) - \cos\left(\dfrac{\pi}{6}\right)}{\Delta x}$

EXERCISE 14-2

Answer each question about continuity.

1. Is the function $f(x) = \begin{cases} \dfrac{x^2 - 25}{x - 5}, & x \neq 5 \\ 10, & x = 5 \end{cases}$ continuous at $x = 5$? Explain.

2. Is the function $g(x) = \begin{cases} \dfrac{x^2 - 16}{x - 4}, & x \neq 4 \\ 7, & x = 4 \end{cases}$ continuous at $x = 4$? Explain.

3. Is the function $p(x) = \dfrac{3x + 1}{x^2 + 4}$ continuous?

EXERCISE 14-3

Compute the derivative of the given function.

1. $g(x) = 5x^8 - 9x^4 + 3$

2. $k(x) = x\sqrt{x^2 + 3}$

3. $p(x) = \dfrac{2x + 3}{5x - 4}$

4. $t(x) = \tan(x)$ [Hint: Use the quotient rule.]

5. $w(x) = \sin^3(x^2)$

6. $v(x) = \dfrac{x\cos(x)}{x^2 - 1}$

EXERCISE 14-4

Write the equation for the indicated tangent line to the graph of the given function.

1. $f(x) = 5x^3 + 4x - 3$ at $x = 2$.

2. $g(x) = 4\sin(2x)$ at $x = \dfrac{5\pi}{6}$.

EXERCISE 14-5

Determine the coordinates of the relative extreme values for the graphs of the given functions.

Flashcard App

1. $b(x) = 3x^4 - 4x^3 - 36x^2 + 4$

2. $c(x) = \dfrac{x}{x^2 + 1}$

Answer Key

Functions

EXERCISE 1-1

1. $D_C = \{-4, -2, 0, 2, 5\}$

2. $R_C = \{3, 5, -9, -1\}$.

3. $C = \{(3, -4), (5, -2), (3, 0), (-9, 2), (-1, 5)\}$

4. C is a function because each input value has a unique output value.

5. C^{-1} is not a function because the input value 3 has two different output values.

6. $f(3) = -2(3)^2 + 50 = -2(9) + 50 = 32$

7. $f(-3) = -2(-3)^2 + 50 = -2(9) + 50 = 32$

8. $g(9) = \dfrac{5(9) + 7}{9 + 4} = \dfrac{45 + 7}{13} = \dfrac{52}{13} = 4$

9. $g(-2) = \dfrac{5(-2) + 7}{-2 + 4} = \dfrac{-10 + 7}{2} = \dfrac{-3}{2}$

10. $h(1) = \sqrt{16 - 7(1)} = \sqrt{9} = 3$

11. $h(-15) = \sqrt{16 - 7(-15)} = \sqrt{16 + 105} = \sqrt{121} = 11$

12. $f(3) + h(-15) = 32 + 11 = 43$

13. $\dfrac{f(3)}{h(1)} = \dfrac{32}{3}$

14. $f(-3) + 3g(9) - 2h(-15) = 32 + 3(4) - 2(11) = 32 + 12 - 22 = 22$

15. All real numbers

16. $x + 4 \neq 0 \Rightarrow x \neq -4$

17. $16 - 7x \geq 0 \Rightarrow x \leq \dfrac{16}{7}$

EXERCISE 1-2

1. $m(\sqrt{36}) = m(6) = \dfrac{3}{10}$

2. $m(8) = \dfrac{5}{12}$

3. $p(n(-6)) = p(\sqrt{16}) = p(4) = 32$

4. $x \leq -2$

5. The domain for $p(x)$ is the set of all real numbers. $n(p(x)) = \sqrt{4 - 4x^2}$, so the domain is the set that satisfies $4 - 4x^2 \geq 0$ and that set is $-1 \leq x \leq 1$.

6. $2(\sqrt{4 - 2x})^2 = 2|4 - 2x|$

7. The cost of production three hours after the business opens

EXERCISE 1-3

1. $y = f(x) = 5 - 7x \Rightarrow x = 5 - 7y \Rightarrow 7y = 5 - x \Rightarrow f^{-1}(x) = \dfrac{5 - x}{7}.$

2. $y = g(x) = \dfrac{5 + 3x}{2x - 7} \Rightarrow x = \dfrac{5 + 3y}{2y - 7} \Rightarrow x(2y - 7) = 5 + 3y \Rightarrow 2xy - 7x$

$= 5 + 3y \Rightarrow 2xy - 3y = 5 + 7x \Rightarrow y(2x - 3) = 5 + 7x \Rightarrow y = \dfrac{5 + 7x}{2x - 3}$

3. $y \neq \dfrac{3}{2}$

4. $y = h(x) = \dfrac{ax + b}{cx + d} \Rightarrow x = \dfrac{ay + b}{cy + d} \Rightarrow x(cy + d) = ay + b$

$\Rightarrow cxy + dx = ay + b.$

$\Rightarrow cxy - ay = b - dx \Rightarrow y(cx - a) = b - dx \Rightarrow f^{-1}(x) = \dfrac{b - dx}{cx - a}$

5. $y \neq \dfrac{a}{c}$

6. The inverse is not a function because it fails the horizontal line test.

Quadratic Functions

EXERCISE 2-1

1. $(4x + 5)(x - 5) = 0 \Rightarrow x = \dfrac{-5}{4}, 5$

2. $(6x - 11)(6x + 11) = 0 \Rightarrow x = \pm\dfrac{11}{6}$

3. $(2x + 7)(4x - 3) = 0 \Rightarrow x = \dfrac{-7}{2}, \dfrac{3}{4}$

4. $3x(16x^2 - 81) = 0 \Rightarrow 3x(4x - 9)(4x + 9) = 0 \Rightarrow x = 0, \pm\dfrac{9}{4}$

5. $(5x + 3)(2x - 9) = 0 \Rightarrow x = \dfrac{-3}{5}, \dfrac{9}{2}$

6. $x = \dfrac{39 \pm \sqrt{39^2 - 4(10)(-26)}}{2(10)} = \dfrac{39 \pm \sqrt{1521 + 1040}}{20} = \dfrac{39 \pm \sqrt{2561}}{20}$;

$-0.580, 4.480$

7. $x = \dfrac{-22 \pm \sqrt{22^2 - 4(8)(-19)}}{2(8)} = \dfrac{-22 \pm \sqrt{484 + 608}}{16} = \dfrac{-22 \pm \sqrt{1092}}{16}$;

$-3.440, 0.690$

8. $x = \dfrac{19 \pm \sqrt{(-19)^2 - 4(5)(13)}}{2(5)} = \dfrac{19 \pm \sqrt{361 - 260}}{10} = \dfrac{19 \pm \sqrt{101}}{10}$;

$0.895, 2.905$

9. $x = \dfrac{-11 \pm \sqrt{(11)^2 - 4(-12)(-2)}}{2(-12)} = \dfrac{-11 \pm \sqrt{121 - 96}}{-24}$

$= \dfrac{-11 \pm \sqrt{25}}{-24} = \dfrac{-11 + 5}{-24}, \dfrac{-11 - 5}{-24} = \dfrac{1}{4}, \dfrac{2}{3}$

10. The common denominator for the three coefficients is 189. Multiply both sides of the equation by 189 to change $\dfrac{1}{3}x^2 + \dfrac{3}{7}x - \dfrac{7}{9} = 0$ to $63x^2 + 81x - 147 = 0$.

$x = \dfrac{-81 \pm \sqrt{(81)^2 - 4(63)(-147)}}{2(63)} = \dfrac{-81 \pm \sqrt{6561 + 37044}}{126}$

$= \dfrac{-81 \pm \sqrt{436}}{126}$; $-2.300, 1.014$

11.

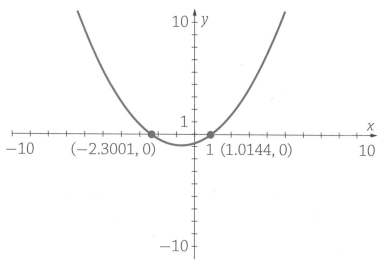

12.

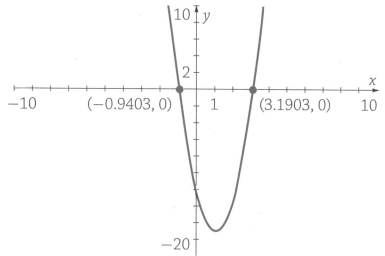

13.

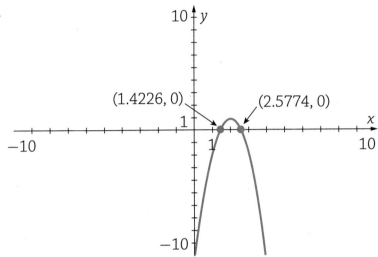

14.

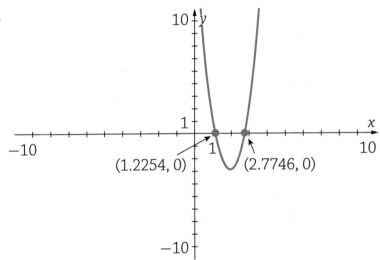

15. No real solution

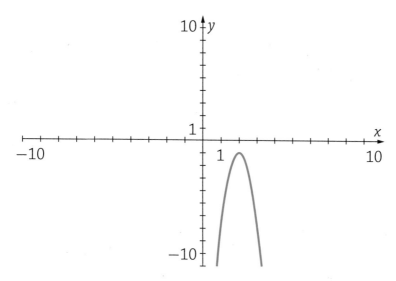

EXERCISE 2-2

1. $(18 + 19i\sqrt{9}\sqrt{2}) + (15 - 10i\sqrt{25}\sqrt{2}) \Rightarrow (18 + 57i\sqrt{2}) + (15 - 50i\sqrt{2})$
   $\Rightarrow 33 + 7i\sqrt{2}$

2. $(81 + 91i\sqrt{16}\sqrt{3}) - (51 - 40i\sqrt{25}\sqrt{3}) \Rightarrow (81 + 364i\sqrt{3}) -$
   $(51 - 200i\sqrt{3}) \Rightarrow 30 + 564i\sqrt{3}$

3. $96 - 32i\sqrt{25}\sqrt{2} + 132i\sqrt{9}\sqrt{2} - 44i^2\sqrt{900} \Rightarrow 96 - 160i\sqrt{2} +$
   $396i\sqrt{3} + (44)(30) \Rightarrow 1416 + 236i\sqrt{2}$

4. $64 + 176i\sqrt{4}\sqrt{6} + 121i^2(\sqrt{24})^2 \Rightarrow 64 + 352i\sqrt{6} - (121)(24)$
   $\Rightarrow 64 + 352i\sqrt{6} - 2904 \Rightarrow -2840 + 352i\sqrt{6}$

5. $17^2 + (2\sqrt{13})^2 \Rightarrow 289 + 52 \Rightarrow 341$

6. $\left(\dfrac{17+2i\sqrt{3}}{8-4i\sqrt{3}}\right)\left(\dfrac{8+4i\sqrt{3}}{8+4i\sqrt{3}}\right) \Rightarrow \dfrac{136+68i\sqrt{3}+16i\sqrt{3}-24}{64+48}$

$\Rightarrow \dfrac{112+84i\sqrt{3}}{112} \Rightarrow 1+\dfrac{3\sqrt{3}}{4}i$

7. $x = (11-6i\sqrt{3})-(13+2i\sqrt{3}) = -2-8i\sqrt{3}$

8. $x = \left(\dfrac{5-2i}{4+3i}\right)\left(\dfrac{4-3i}{4-3i}\right) = \dfrac{20-15i-8i+6i^2}{16+9} = \dfrac{20-23i-6}{25}$

$= \dfrac{14}{25}-\dfrac{23}{25}i$

9. $x = (5-8i\sqrt{2})(2+3i\sqrt{2}) = 10+15i\sqrt{2}-16i\sqrt{2}-12i^2(4)$

$= 10-i\sqrt{2}+48 = 58-i\sqrt{2}$

10. $x = \left(\dfrac{2+3i\sqrt{2}}{5-8i\sqrt{2}}\right)\left(\dfrac{5+8i\sqrt{2}}{5+8i\sqrt{2}}\right) = \dfrac{10+16i\sqrt{2}+15i\sqrt{2}+24i^2(2)}{25+128}$

$= \dfrac{-38}{153}+\dfrac{31\sqrt{2}}{153}i$

EXERCISE 2-3

1. Axis of symmetry: $x = \dfrac{12}{2(-2)} = -3$; Vertex: $f(-3) = 37$; Range: $y \le 37$

2. Vertex: $(-6, 10)$; Range: $y \ge 10$

3. Axis of symmetry: $x = \dfrac{12}{2\left(\dfrac{3}{5}\right)} = 10$; Vertex: $g(10) = -41$; Range: $y \ge -41$

4. Vertex: $(6, 7)$; Range: $y \ge 7$

5. Axis of symmetry: $x = \dfrac{-21}{2\left(\dfrac{-3}{7}\right)} = \dfrac{49}{2}$; Vertex: $g\left(\dfrac{49}{2}\right) = \dfrac{1465}{4} = 366.25$;

Range: $y \le 366.25$

6. Axis of symmetry: $g = \dfrac{-6}{2(-0.02)} = 150$; Vertex: $P(15) = 425$; The maximum daily profit is \$425.

7. Axis of symmetry: $7 = \dfrac{-96}{2(-16)} = 3$; Vertex: $h(3) = 150$; the maximum height of the ball is 150 feet.

8. 45 by 45 feet

9. $l + 2w = 180 \Rightarrow l = 180 - 2w$; $A = lw \Rightarrow A = (180 - 2w)w =$

$-2w^2 + 180w$; the dimensions of the rectangle are 45 by 90 feet.

10. $5l + 2(l + 2w) = 200 \Rightarrow 7l + 4w = 200 \Rightarrow 4w = 200 - 7l$

$\Rightarrow w = 50 - \dfrac{7}{4}l$; $A = l\left(50 - \dfrac{7}{4}l\right) = \dfrac{-7}{4}l^2 + 50l$; axis of symmetry:

$l = \dfrac{-50}{2\left(\dfrac{-7}{4}\right)} = \dfrac{100}{7}$; $w = 50 - \dfrac{7}{4}\left(\dfrac{100}{7}\right) = 25$; the dimensions of the

garden with maximum area are $\dfrac{100}{7}$ by 25 feet.

Polynomial Functions

EXERCISE 3-1

1. $(2x-1)(64x^6 + 32x^5 + 16x^4 + 8x^3 + 4x^2 + 2x + 1)$

2. If $f(x) = 2x^5 - 3x^4 + 7x^2 + 5x - 11$ then $f(1) = 0$. Therefore, by the Remainder theorem, the remainder is 0.

3. $x - 1$ is a factor of $2x^5 - 3x^4 + 7x^2 + 5x - 11$ because the remainder is 0 when $2x^5 - 3x^4 + 7x^2 + 5x - 11$ is divided by $x - 1$.

4. $2x^4 - x^3 - x^2 + 6x + 11$

$$
\begin{array}{r|rrrrrr}
1 & 2 & -3 & 0 & 7 & 5 & -11 \\
  &   & 2 & -1 & -1 & 6 & 11 \\
\hline
  & 2 & -1 & -1 & 6 & 11 & 0
\end{array}
$$

5. $2x^3 + 3x^2 - 9x - 10$

$$
\begin{array}{r|rrrrr}
\dfrac{4}{3} & 6 & 1 & -39 & 6 & 40 \\
  &   & 8 & 12 & -36 & -40 \\
\hline
  & 6 & 9 & -27 & -30 & 0
\end{array}
$$

$$\frac{6x^4 + x^3 - 39x^2 + 6x + 40}{3x - 4} = \frac{6x^4 + x^3 - 39x^2 + 6x + 40}{3\left(x - \dfrac{4}{3}\right)} =$$

$$\frac{6x^3 + 9x^2 - 27x - 30}{3} = 2x^3 + 3x^2 - 9x - 10$$

6. $x^2 - x - 2$

$$\frac{-5}{2}\bigg|\ \ 2\ \ \ 3\ \ -9\ \ -10$$
$$\phantom{\frac{-5}{2}\bigg|\ \ }\ -5\ \ \ \ 5\ \ \ \ \ 10$$
$$\overline{\phantom{\frac{-5}{2}\bigg|\ }2\ -2\ \ -4\ \ \ \ \ 0}$$

$$\frac{2x^3 + 3x^2 - 9x - 10}{2x + 5} = \frac{2x^3 + 3x^2 - 9x - 10}{2\left(x + \frac{5}{2}\right)} = \frac{2x^2 - 2x - 4}{2}$$

$$= x^2 - x - 2$$

7. $6x^4 + x^3 - 39x^2 + 6x + 40 = (3x - 4)(2x + 5)(x^2 - x - 2)$
$= (3x - 4)(2x + 5)(x - 2)(x + 1)$

8. Odd degree with positive coefficient: $x \to -\infty \Rightarrow y \to -\infty$
and $x \to \infty \Rightarrow y \to \infty$

9. Even degree with negative coefficient: $x \to -\infty \Rightarrow y \to -\infty$
and $x \to \infty \Rightarrow y \to -\infty$

10. Odd degree with negative coefficient: $x \to -\infty \Rightarrow y \to \infty$
and $x \to \infty \Rightarrow y \to -\infty$

EXERCISE 3-2

1. $(2x + 7)^2(x - 4)(x - 7)$

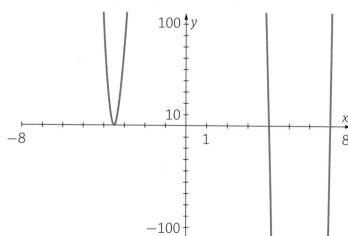

2. $(x + 4)(x - 2)^3(x - 5)$

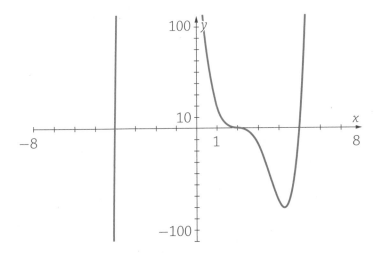

3. $(2x + 5)(3x - 2)(x^2 + x + 4)$

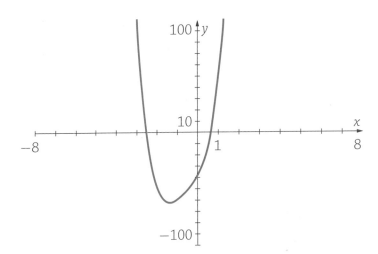

Use synthetic division to reduce the polynomial.

$$\begin{array}{r|rrrrr} \dfrac{-5}{2} & 6 & 17 & 25 & 34 & -40 \\ & & -15 & -5 & -50 & 40 \\ \hline & 6 & 2 & 20 & 16 & 0 \end{array}$$

Divide coefficients by 2.

$$
\dfrac{2}{3}\Big|\quad 3\quad 1\quad 10\quad -8
$$

$$
\underline{\qquad\quad 2\quad 2\quad\ \ 8\ \ }
$$

$$
3\ 3\quad 12\quad\ \ 0
$$

Divide coefficients by 3.

4. $(x+2)^2(x-2)^2(x^2+4)$

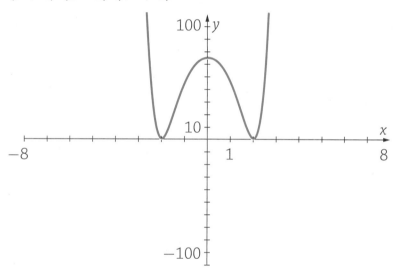

$$
\begin{array}{r|rrrrrrr}
-2 & 1 & 0 & -4 & 0 & -16 & 0 & 64 \\
 & & -2 & 4 & 0 & 0 & 32 & -64 \\
\hline
-2 & 1 & -2 & 0 & 0 & -16 & 32 & 0 \\
 & & -2 & 8 & -16 & 32 & -32 & \\
\hline
2 & 1 & -4 & 8 & -16 & 16 & 0 & \\
 & & 2 & -4 & 8 & -16 & & \\
\hline
2 & 1 & -2 & 4 & -8 & 0 & & \\
 & & 2 & 0 & 8 & & & \\
\hline
 & 1 & 0 & 4 & 0 & & &
\end{array}
$$

5. $(x + 3)(2x + 1)(4x - 3)(x^2 + 3x + 4)$

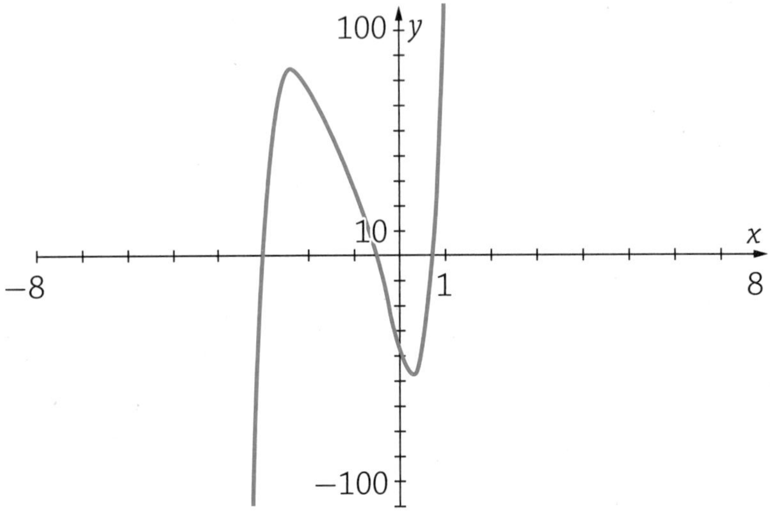

$$
\begin{array}{r|rrrrrr}
-3 & 8 & 46 & 89 & 52 & -63 & -36 \\
  &   & -24 & -66 & -69 & 51 & 36 \\
\hline
-\dfrac{1}{2} & 8 & 22 & 23 & -17 & -12 & 0 \\
  &   & -4 & -9 & -7 & 12 \\
\hline
\dfrac{3}{4} & 8 & 18 & 14 & -24 & 0 \\
  &   & 6 & 18 & 24 \\
\hline
  & 8 & 24 & 32 & 0
\end{array}
$$

Divide coefficients by 8.

6. $x = \dfrac{-7}{2}, 4, 7$

7. $x = -4, 2, 5$

8. $x = \dfrac{-5}{2}, \dfrac{2}{3}, \dfrac{-1}{2} \pm \dfrac{\sqrt{15}}{2} i$

9. $x = -2, 2, \pm 2i$

10. $x = -3, \dfrac{-1}{2}, \dfrac{3}{4}, \dfrac{-3}{2} \pm \dfrac{\sqrt{7}}{2}i$ c

11. $4 < x < 7$

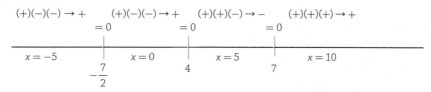

$(+)(-)(-) \to +$ $\quad$ $(+)(-)(-) \to +$ $\quad$ $(+)(+)(-) \to -$ $\quad$ $(+)(+)(+) \to +$

$\qquad = 0 \qquad\qquad = 0 \qquad\qquad = 0$

$x = -5 \qquad\quad x = 0 \qquad\quad x = 5 \qquad\quad x = 10$

$\qquad -\dfrac{7}{2} \qquad\qquad 4 \qquad\qquad 7$

12. $-4 < x < 2$ or $x \geq 5$

$(-)(-)(-) \to -$ $\quad$ $(+)(-)(-) \to +$ $\quad$ $(+)(+)(-) \to -$ $\quad$ $(+)(+)(+) \to +$

$\qquad = 0 \qquad\qquad = 0 \qquad\qquad = 0$

$x = -5 \qquad\quad x = 0 \qquad\quad x = 4 \qquad\quad x = 10$

$\qquad -4 \qquad\qquad 2 \qquad\qquad 5$

13. $\dfrac{-5}{2} \leq x \leq \dfrac{2}{3}$

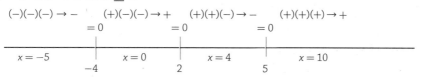

$(-)(-)(+) \to +$ $\quad$ $(+)(-)(+) \to -$ $\quad$ $(+)(+)(+) \to +$

$\qquad = 0 \qquad\qquad = 0$

$x = -5 \qquad\quad x = 0 \qquad\quad x = 4$

$\qquad -\dfrac{5}{2} \qquad\qquad \dfrac{2}{3}$

14. All real numbers

$(+)(+)(+) \to +$ $\quad$ $(+)(+)(+) \to +$ $\quad$ $(+)(+)(+) \to +$

$\qquad = 0 \qquad\qquad = 0$

$x = -5 \qquad\quad x = 0 \qquad\quad x = 4$

$\qquad -2 \qquad\qquad 2$

15. $-3 < x < \dfrac{-1}{2}$ or $x > \dfrac{3}{4}$

$$(-)(-)(-)(+) \rightarrow - \quad (+)(-)(-)(+) \rightarrow + \quad (+)(+)(-)(+) \rightarrow - \quad (+)(+)(+)(+) \rightarrow +$$
$$= 0 \qquad\qquad = 0 \qquad\qquad = 0$$

$x = -5$	$x = -2$	$x = 0$	$x = 10$
-3	$\dfrac{-1}{2}$		$\dfrac{3}{4}$

16. q: 1, 2, 3, 5, 6, 10, 15, 30; p: 1, 2, 3, 4, 6, 8, 12, 24

$$\frac{p}{q} = \pm \left\{ \begin{array}{l} 1, 2, 3, 4, 6, 8, 12, 24, \dfrac{1}{2}, \dfrac{3}{2}, \dfrac{1}{3}, \dfrac{2}{3}, \dfrac{4}{3}, \dfrac{8}{3}, \dfrac{1}{5}, \dfrac{2}{5}, \dfrac{3}{5}, \dfrac{4}{5}, \dfrac{6}{5}, \dfrac{8}{5}, \dfrac{12}{5}, \dfrac{24}{5}, \\ \dfrac{1}{6}, \dfrac{1}{10}, \dfrac{3}{10}, \dfrac{1}{15}, \dfrac{2}{15}, \dfrac{4}{15}, \dfrac{8}{15}, \dfrac{1}{30} \end{array} \right\}$$

17. Local maximum: 11.4; local minimum: −11.8

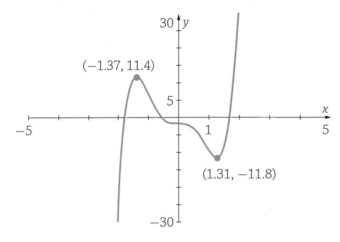

18. Local minimum: −2.67; no local maximum

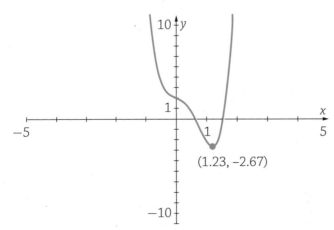

19. Local minimum: 4; local maximum: 12

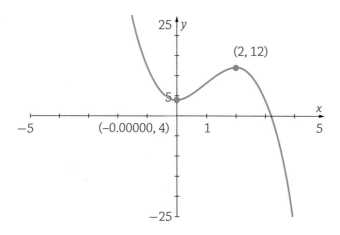

20. Local maximum: 39; local minimum: 3

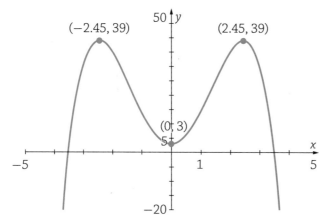

Rational Functions

EXERCISE 4-1

1. y intercept: 1; z intercepts: $z^2 - 8z - 9 = 0 \Rightarrow z = -1, 9$

2. y intercept: 5; v intercepts: $v + 5 = 0 \Rightarrow v = -5$

3. y intercept: -1; p intercepts: $d(p) = \dfrac{4p^2 - 4p + 1}{4p^2 - 1} = \dfrac{(2p-1)^2}{(2p-1)(2p+1)} = \cdot$

$\dfrac{2p-1}{2p+1} \Rightarrow$ none because $p = \dfrac{1}{2}$ is not in the domain.

4. y intercept: 0; t intercept: $t^3 - 4t = 0 \Rightarrow t = 0, \pm 2$

5. y intercept: 3; a intercept: $w(a) = \dfrac{6a - 12}{a^2 - 4} = \dfrac{6(a-2)}{(a-2)(a+2)} = \dfrac{6}{a+2} \Rightarrow$

none because $a = 2$ is not in the domain.

6. Infinite discontinuities at $z = -3$ and at $z = 3$.

$z \to -3^- \Rightarrow y \to \infty$ and $z \to -3^+ \Rightarrow y \to -\infty$; $z \to 3^- \Rightarrow y \to \infty$

and $z \to 3^+ \Rightarrow y \to -\infty$

$\dfrac{(-)(-)}{(-)(-)} \to +$ $\dfrac{(-)(-)}{(+)(-)} \to -\ = 0$ $\dfrac{(+)(-)}{(+)(-)} \to +$ $\dfrac{(+)(-)}{(+)(+)} \to -\ = 0$ $\dfrac{(+)(+)}{(+)(+)} \to +$

$z = -5$	$z = -2$	$z = 0$	$z = 4$	$z = 10$
-3	-1	3	9	

7. There are no discontinuities.

8. Infinite discontinuity at $p = \dfrac{-1}{2}$ and a point discontinuity at $\left(\dfrac{1}{2}, 0\right)$.

$p \to -\dfrac{1}{2}^- \Rightarrow y \to -\infty$ and $p \to -\dfrac{1}{2}^+ \Rightarrow y \to \infty$

$\dfrac{(-)}{(+)} \to -$ $\dfrac{(+)}{(+)} \to +$

$p = 0$	$p = 1$
$\dfrac{-1}{2}$	

9. Infinite discontinuities at $t = -5$ and $t = 5$.

$t \rightarrow -5^- \Rightarrow y \rightarrow -\infty$ and $t \rightarrow -5^+ \Rightarrow y \rightarrow \infty$; $t \rightarrow 5^- \Rightarrow y \rightarrow -\infty$

and $t \rightarrow 5^+ \Rightarrow y \rightarrow \infty$

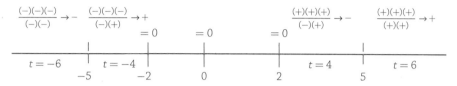

10. Infinite discontinuity at $a = -2$ and a point discontinuity at $\left(2, \dfrac{3}{2}\right)$.

$a \rightarrow -2^- \Rightarrow y \rightarrow -\infty$ and $a \rightarrow -2^+ \Rightarrow y \rightarrow \infty$

EXERCISE 4-2

1. $x \rightarrow \pm\infty \Rightarrow y \rightarrow 1$

2. $x \rightarrow \pm\infty \Rightarrow y \rightarrow 2$

3. $x \rightarrow \pm\infty \Rightarrow y \rightarrow \dfrac{2}{3}$

4. $x \rightarrow \pm\infty \Rightarrow y \rightarrow 0$

5. $x \rightarrow -\infty \Rightarrow y \rightarrow -\infty$; $x \rightarrow \infty \Rightarrow y \rightarrow \infty$

6. Intercepts: $\left(0, \dfrac{25}{4}\right)$; $(\pm 5, 0)$ Vertical asymptotes: $x = \pm 2$ Horizontal asymptote: $y = 1$

Signs analysis:

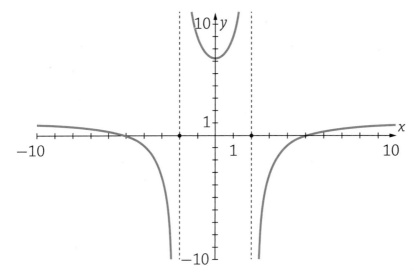

7. Intercepts: $\left(0, \dfrac{3}{2}\right); \left(\dfrac{-3}{2}, 0\right), (2, 0)$ Vertical asymptotes: $x = -4; x = 1$

Horizontal asymptote: $y = 2$

Signs analysis:

$\dfrac{(-)(-)}{(-)(-)} \to +$ $\dfrac{(+)(-)}{(-)(-)} \to -$ $\dfrac{(+)(-)}{(+)(-)} \to +$ $\dfrac{(+)(-)}{(+)(+)} \to -$ $\dfrac{(+)(+)}{(+)(+)} \to +$

 $= 0$ $= 0$

$x = -5$ $x = -2$ $x = 0$ $x = 1.5$ $x = 5$

 -4 $\dfrac{-4}{3}$ 1 2

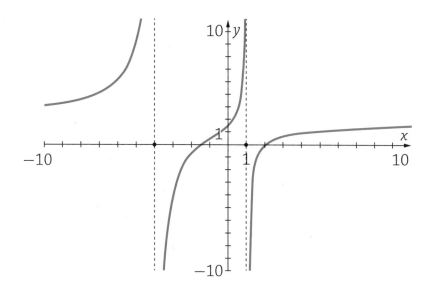

8. Intercepts: $\left(0, \dfrac{15}{8}\right)$; $\left(\dfrac{-5}{2}, 0\right)$, $(3, 0)$ Vertical asymptotes: $x = \dfrac{-4}{3}$, $x = 2$

Horizontal asymptote: $y = \dfrac{2}{3}$

Signs analysis:

$$\dfrac{(-)(-)}{(-)(-)} \to +$$

$$= 0$$

$$\dfrac{(+)(-)}{(-)(-)} \to -$$

$$\dfrac{(+)(-)}{(+)(-)} \to +$$

$$\dfrac{(+)(-)}{(+)(+)} \to -$$

$$= 0$$

$$\dfrac{(+)(+)}{(+)(+)} \to +$$

$x = -6$ $x = -3$ $x = 0$ $x = 3$ $x = 6$

$\dfrac{-5}{2}$ $\dfrac{-4}{3}$ 2 3

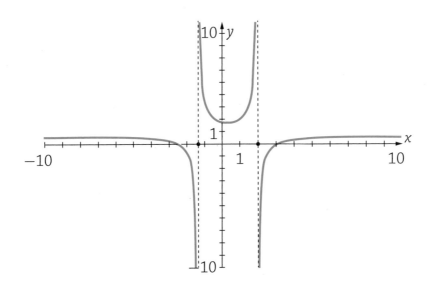

9. Reduced expression: $p(x) = x - 1 + \dfrac{2}{x-2}$

$$
\begin{array}{r|rrr}
2\rfloor & 1 & -3 & 4 \\
  &   & 2 & -2 \\
\hline
  & 1 & -1 & 2
\end{array}
$$

Intercepts: $(0, -2)$ Vertical asymptote: $x = 2$ Oblique asymptote: $y = x - 1$

Signs analysis:

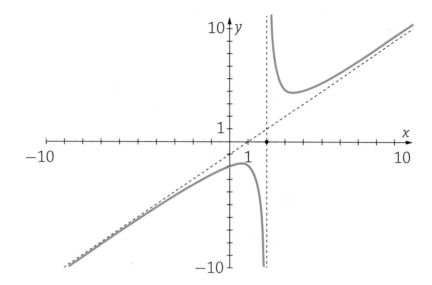

10. $r(x) = \dfrac{(x-1)^2(x^2+16)}{x^4-16} = \dfrac{(x-1)^2(x^2+16)}{(x^2+4)(x^2-4)}$

Intercepts: $(0, -1)$, $(1, 0)$ Vertical asymptotes: $x = \pm 2$ Horizontal asymptote: $y = 1$

Signs analysis:

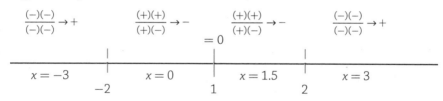

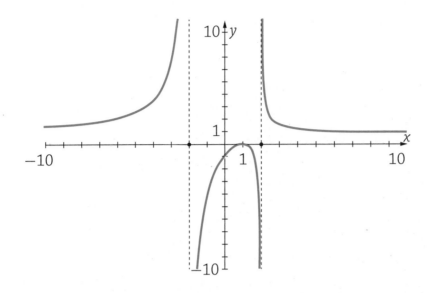

11. Local minimum: −1.25; local maximum: 1.25

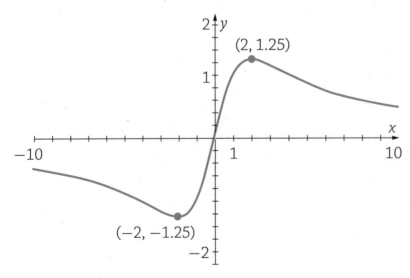

12. Local minimum: 3.83; local maximum: −1.83

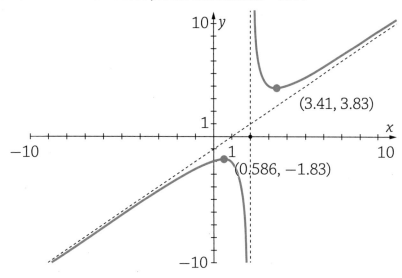

(3.41, 3.83)

(0.586, −1.83)

13. Local maximum: 0; no local minimum

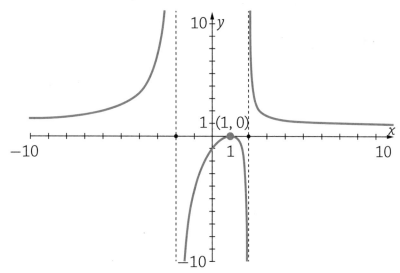

(1, 0)

14. Radius: 4.01 cm   Height: 9.36 cm

Volume: $V = \pi r^2 h = 473.18 \Rightarrow h = \dfrac{473.18}{\pi r^2}$

Cost: $C = 0.02(\pi r^2) + 0.015(\pi r^2 + 2\pi rh) = 0.035\pi r^2 + 0.03\pi rh$

$\Rightarrow C = 0.035\pi r^2 + 0.03\pi r\left(\dfrac{473.18}{\pi r^2}\right) \Rightarrow C = 0.035\pi r^2 + \dfrac{14.1854}{r}$

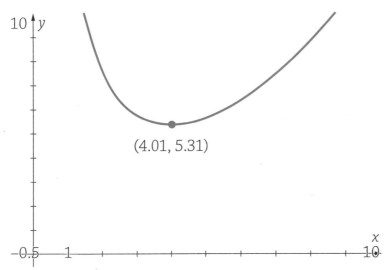

(4.01, 5.31)

15. Length $= 37.9$ ft; width $= 47.4$ ft.

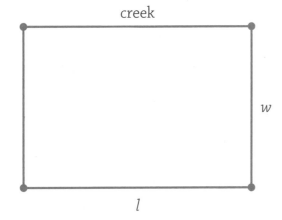

creek

w

l

Area: $lw = 1800 \Rightarrow w = \dfrac{1800}{l}$

Cost: $C = 6l + 4(l + 2w) = 10l + 8w \Rightarrow C = 10l + \dfrac{14400}{l}$

Graph:

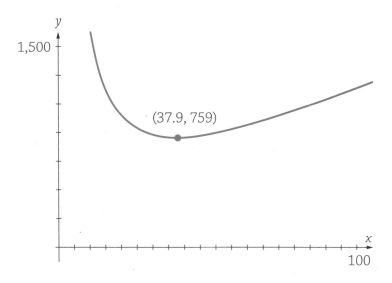

(37.9, 759)

Conic Sections

EXERCISES 5-1

1. The radius is $6 \Rightarrow (x-5)^2 + (y-6)^2 = 36$

2. The center of the circle is (4, 11) making the equation of the form $(x-4)^2 + (y-11)^2 = r^2$. Substituting the coordinates of one of the points in the circle gives $(x-4)^2 + (y-11)^2 = 80$.

3. $(0, -1), (0, 11), (-2 + \sqrt{15}, 0), (-2 - \sqrt{15}, 0)$

   Set $x = 0 \Rightarrow 4 + (y-5)^2 = 40 \Rightarrow (y-5)^2 = 36 \Rightarrow y = -1, 11$

Set $y = 0 \Rightarrow (x+2)^2 + 25 = 40 \Rightarrow (x+2)^2 = 15 \Rightarrow x = -2 \pm \sqrt{15}$

EXERCISES 5-2

1. Vertex (4, 4); $p = 2$; $y - 4 = \dfrac{1}{8}(x-4)^2$

2. Vertex (3, 6); $p = 1$; $x - 3 = \dfrac{1}{4}(y-6)^2$

3. $p = 2$; $x + 3 = \dfrac{-1}{8}(y-2)^2$

4. Vertex (3, −6); $p = -5 \Rightarrow$ Focus (3, −11); directrix: $y = -1$

5. Vertex (1, 4); $p = 3 \Rightarrow$ Focus (4, 4); directrix: $x = -2$

EXERCISES 5-3

1. Center (−1, 5); $c = 2$, $a = 5 \Rightarrow b^2 = 21$; $\dfrac{(x+1)^2}{25} + \dfrac{(y-5)^2}{21} = 1$

2. Center (2, 4); $a = 8$; $b = 5$; $\dfrac{(x-2)^2}{64} + \dfrac{(y-4)^2}{25} = 1$

3. $c = 5$; $a = 8 \Rightarrow b^2 = 39$; $\dfrac{(x-5)^2}{64} + \dfrac{(y-6)^2}{39} = 1$

EXERCISES 5-4

1. $\dfrac{(x+1)^2}{4} + \dfrac{(y-1)^2}{1} = 1 \Rightarrow$ Center (−1, 1); foci $(-1 \pm \sqrt{3}, 1)$; major

   axis: (−3, 1), (1, 1); minor axis: (−1, 0), (−1, 2); $e = \dfrac{\sqrt{3}}{2}$

2. $\dfrac{(x+2)^2}{4} + \dfrac{(y-1)^2}{9} = 1 \Rightarrow$ Center (−2, 1); foci $(-2, 2 \pm \sqrt{5})$; major

   axis (−2, −2), (−2, 4); minor axis: (−4, 1), (0, 1); $e = \dfrac{\sqrt{5}}{3}$

3. $\dfrac{(x+5)^2}{100} + \dfrac{(y-4)^2}{36} = 1 \Rightarrow$ Center $(-5, 4)$; foci $(-3, 4)$, $(13, 4)$; major

   axis: $(-15, 4)$, $(5, 4)$; minor axis: $(-5, -2)$, $(-5, 10)$; $e = \dfrac{4}{5}$

4. 15 feet high in the semi-major axis, so $a = 15$; 20 feet wide is the minor axis, so $b = 10$. Identifying the middle of the road under the arch as the origin, the equation for the ellipse is $\dfrac{x^2}{100} + \dfrac{y^2}{225} = 1$. If the driver puts the center of the truck in the middle of the road, the edge of the truck will be 6 feet from the center. The maximum height of the truck at that point will be the positive solution to the equation $\dfrac{36}{100} + \dfrac{y^2}{225} = 1$, or 12 feet. So, the truck should be a little less than 12 feet tall so as to be able to safely pass through the arch.

EXERCISES 5-5

1. The major axis is 97 feet, so $a = 48.5$. The minor axis is 46 feet, making $b = 23$. Placing the origin at the center of the room, the equation is $\dfrac{x^2}{48.5^2} + \dfrac{y^2}{23^2} = 1$.

2. The value of c is $\sqrt{48.5^2 - 23^2} = 42.70$, so the two foci are 85.4 feet apart.

3. The length of each focal chord is $\dfrac{2b^2}{a} = \dfrac{2(23)^2}{48.5} = 21.8$ feet.

EXERCISES 5-6

1. $\dfrac{y^2}{4} - \dfrac{(x-1)^2}{9} = 1$; transverse axis is vertical and is of length $4 \Rightarrow a = 2$; $b = 3$ and center is at $(1, 0)$.

2. $\dfrac{(x+2)^2}{36} - \dfrac{(y-2)^2}{28} = 1$; transverse axis is horizontal; $a = 6$; $c = 8$

3. $\dfrac{(y-2)^2}{16} - \dfrac{(x-4)^2}{9} = 1$; transverse axis is vertical; $a = 4$; slope of

asymptote $= \dfrac{4}{3} \Rightarrow b = 3$.

EXERCISES 5-7

1. Center: $(-3, 1)$; vertices: $(-3 \pm \sqrt{2}, 1)$; foci: $(-3 \pm \sqrt{3}, 1)$; endpoints of

the focal chords $(-3 \pm \sqrt{3}, 1 \pm \sqrt{2})$; asymptotes: $y - 1 = \pm \dfrac{1}{\sqrt{2}}(x + 3)$;

$e = \sqrt{\dfrac{3}{2}}$

$\dfrac{(x+3)^2}{2} - \dfrac{(y-1)^2}{1} = 1$; transverse axis is horizontal; $a = \sqrt{2}$;

$b = 1$; $c = \sqrt{3}$

2. Center: $(-6, -1)$; vertices: $(-6, -3)$, $(-6, 1)$; foci: $(-6, -1 \pm \sqrt{53})$;

endpoints of the focal chords: $\left(-6 \pm \dfrac{49}{2}, -1 \pm \sqrt{53}\right) =$

$\left(\dfrac{-61}{2}, -1 \pm \sqrt{53}\right), \left(\dfrac{37}{2}, -1 \pm \sqrt{53}\right)$; asymptotes: $y + 1 = \pm \dfrac{2}{7}(x + 6)$;

$e = \dfrac{\sqrt{53}}{7}$

$\dfrac{(y+1)^2}{4} - \dfrac{(x+6)^2}{49} = 1$; transverse axis is vertical; $a = 2$;

$b = 7$; $c = \sqrt{53}$

3. Center $(-2, -2)$; vertices: $(-4, -2)$, $(0, -2)$; foci: $(-2 \pm \sqrt{13}, -2)$;

endpoints of the focal chord: $\left(-2 \pm \sqrt{13}, -2 \pm \dfrac{9}{2}\right) = \left(-2 \pm \sqrt{13}, \dfrac{-13}{2}\right)$,

$$\left(-2 \pm \sqrt{13}, \frac{5}{2}\right); e = \frac{\sqrt{13}}{3}$$

$$\frac{(x+2)^2}{4} - \frac{(y+2)^2}{9} = 1; \text{ transverse axis is horizontal; } a = 2;$$

$$b = 3; c = \sqrt{13}$$

EXERCISES 5-8

1. $(1, 0), (-1, 0)$

   $$y = x^2 - 1 \Rightarrow x^2 = y + 1 \Rightarrow 2(y+1) + y^2 = 2 \Rightarrow 2y^2 + 2y = 0$$

   $$\Rightarrow y(y+2) = 0 \Rightarrow y = 0, -2$$

   $$y = 0 \Rightarrow x^2 = 1 \Rightarrow x = \pm 1; \ y = -2 \Rightarrow x^2 = -1 \Rightarrow \text{no solution}$$

2. $\left(\dfrac{-3\sqrt{3}}{\sqrt{7}}, \dfrac{8}{\sqrt{7}}\right), \left(\dfrac{-3\sqrt{3}}{\sqrt{7}}, \dfrac{-8}{\sqrt{7}}\right), \left(\dfrac{3\sqrt{3}}{\sqrt{7}}, \dfrac{8}{\sqrt{7}}\right), \left(\dfrac{3\sqrt{3}}{\sqrt{7}}, \dfrac{-8}{\sqrt{7}}\right)$

$3y^2 - 4x^2 = 12$	$4(3y^2 - 4x^2 = 12)$	$12y^2 - 16x^2 = 48$
$16x^2 + 9y^2 = 144$	$16x^2 + 9y^2 = 144$	$9y^2 + 16x^2 = 144$

   $$\Rightarrow 21y^2 = 192 \Rightarrow y = \pm\sqrt{\frac{192}{21}} = \pm\sqrt{\frac{64}{7}} = \pm\frac{8}{\sqrt{7}}$$

   $$y^2 = \frac{64}{7} \Rightarrow 16x^2 + 9\left(\frac{64}{7}\right) = 144 \Rightarrow 16x^2 = \frac{432}{7} \Rightarrow x^2 = \frac{27}{7}$$

   $$\Rightarrow x = \pm\frac{3\sqrt{3}}{\sqrt{7}}$$

3. $(-3, -4), (-4, -3), (4, 3), (3, 4)$

   $$x^2 + \left(\frac{12}{x}\right)^2 = 25 \Rightarrow x^2 + \frac{144}{x^2} = 25 \Rightarrow x^4 - 25x^2 + 144 = 0$$

   $$\Rightarrow (x^2 - 16)(x^2 - 9) = 0 \Rightarrow x = \pm 4, \pm 3$$

4. The slope of the tangent line is $\dfrac{-12}{5}$, so the line containing the center

of the circle(s) and the point of tangency must be $\dfrac{5}{12}$. The equation

of the line with slope $\dfrac{5}{12}$ passing through the point (24, 10) is

$y - 10 = \dfrac{5}{12}(x - 24)$ or $y = \dfrac{5}{12}x$. If the coordinates for the center of the

circle are (h, k), then it must be true that $k = \dfrac{5}{12}h$ and $(24 - h)^2 +$

$(10 - k)^2 = 169$. Substitute for k to get $(24 - h)^2 + \left(10 - \dfrac{5h}{12}\right)^2 = 169$.

Solve this quadratic equation.

$$(24 - h)^2 + \left(10 - \frac{5h}{12}\right)^2 = 169$$

$$(h - 24)^2 + \left(\frac{-5}{12}(h - 24)\right)^2 = 169$$

$$(h - 24)^2 + \frac{25}{144}(h - 24)^2 = 169$$

$$\frac{169}{144}(h - 24)^2 = 169$$

Remove fractions and combine like terms.

$$(h - 24)^2 = 144$$
$$h - 24 = \pm 12$$
$$h = 12, 36$$

When $h = 12$, $k = 5$, giving the equation $(x - 12)^2 + (y - 5)^2 = 169$.

When $h = 36$, $k = 15$, giving the equation $(x - 36)^2 + (y - 15)^2 = 169$.

6

Exponential and Logarithmic Functions

EXERCISES 6-1

1. $\log_b(9 \times 5) = \log_b(3^2) + \log_b(5) = 2\log_b(3) + \log_b(5) = 2y + z$

2. $\log_b(12^2) = 2\log_b(4 \times 3) = 2(\log_b(2^2) + \log_b(3))$
$$= 2(2\log_b(2) + \log_b(3)) = 2(2x + y) = 4x + 2y$$

3. $\log_b(10^2) = 2\log_b(2 \times 5) = 2(\log_b(2) + \log_b(5)) = 2x + 2z$

4. $\log_{10}(3) = \dfrac{\log_b(3)}{\log_b(10)} = \dfrac{\log_b(3)}{\log_b(2) + \log_b(5)} = \dfrac{y}{x + z}$

5. $\log_3(10) = \dfrac{x + z}{y}$

6. $\dfrac{5}{4}$

   $3^{4x} = 3^5$

7. $\dfrac{17}{6}$

   $(2^2)^{3x-5} = 2^7$

8. $\dfrac{5}{3}$

   $(27^{x-1}) = 9 \Rightarrow 3^{3x-3} = 3^2$

9. -1

   $6^{x+4} = 216 = 6^3$

10. $\dfrac{37}{23}$

$(2^3)^{3-5x} = (2^4)^{2x-7} \Rightarrow 9 - 15x = 8x - 28$

11. 17

$5^{2(x+4)} = 5^{3(x-3)}$

12. 8.143

$x = \dfrac{\ln(78)}{\ln(4)} + 5$

13. $12^{2x-1} = 1000$

$2x - 1 = \dfrac{\ln(1000)}{\ln(12)} \Rightarrow x = \dfrac{\dfrac{\ln(1000)}{\ln(12)} + 1}{2}$

14. 1.672

$x = \dfrac{\ln(178)}{3.1}$

15. 0.198

$x = \dfrac{\ln(0.946)}{-0.28}$

16. −0.500

$x = \dfrac{\ln\left(\dfrac{89.6 - 85}{4.1}\right)}{-.23}$

17. 5.517

$x = \dfrac{\ln\left(\dfrac{58 - 49.6}{4.1}\right)}{0.13}$

18. $\dfrac{-5}{3}, 1$

$$7^{3x^2+2x} = 7^5 \Rightarrow 3x^2 + 2x - 5 = 0 \Rightarrow (3x+5)(x-1) = 0$$

19. $\dfrac{3}{2}, -5$

$$(2^{2x^2})(2^{15}) = 2^{-7x} \Rightarrow 2x^2 + 7x - 15 = 0 \Rightarrow (2x-3)(x+5) = 0$$

20. 132

$$x - 4 = 2^7$$

21. 3821.203

$$x = \dfrac{8^{4.3} - 1}{2}$$

22. 0.627

$$x = \dfrac{4 - e^{-0.145}}{5}$$

23. 3

$$\log_2((2x+2)(5x+1)) = 7 \Rightarrow 10x^2 + 12x + 2 = 2^7$$

$$\Rightarrow 10x^2 + 12x - 126 = 0 \Rightarrow 5x^2 + 6x - 63 = 0$$

$$\Rightarrow (5x + 21)(x - 3) = 0 \Rightarrow \dfrac{-21}{5}, 3 \, ; \text{Reject } \dfrac{-21}{5} \text{ because it is not in the}$$
domain.

24. 3

$$\log((x+1)(9x-2)) = 2 \Rightarrow 9x^2 + 7x - 2 = 10^2$$

$$\Rightarrow 9x^2 + 7x - 102 = 0 \Rightarrow (9x + 34)(x - 3) = 0$$

Reject $\dfrac{-34}{9}$ because it is not in the domain.

25. $\dfrac{5}{2}, 10$

$$\log_5\left(\dfrac{2x^2}{x-2}\right) = 2 \Rightarrow \dfrac{2x^2}{x-2} = 5^2 \Rightarrow 2x^2 = 25x - 50$$

$$\Rightarrow 2x^2 - 25x + 50 = 0 \Rightarrow (2x - 5)(x - 10) = 0$$

EXERCISE 6-2

1. $A = 2500\left(1 + \dfrac{.02}{4}\right)^{(10)(4)} = 3051.99$

2. $A = 2500e^{(.02)10} = 2500e^{(.2)} = 3053.51$

3. $2500 = P\left(1 + \dfrac{.036}{4}\right)^{20} \Rightarrow P = \dfrac{2500}{\left(1 + \dfrac{.036}{4}\right)^{20}} = 2089.86$

4. $2500 = 2000e^{5r} \Rightarrow r = \dfrac{\ln\left(\dfrac{5}{4}\right)}{5} = 0.4463 \Rightarrow r = 4.463\%$

5. 2025

$$600 = 540(1.021)^t \Rightarrow t = \dfrac{\ln\left(\dfrac{600}{540}\right)}{\ln(1.021)} = 5.070$$

6. 2064

$$15 = \dfrac{20}{1 + e^{-0.02t}} \Rightarrow 1 + e^{-0.02t} = \dfrac{20}{15} = \dfrac{4}{3} \Rightarrow e^{-0.02t} = \dfrac{1}{3}$$

$$\Rightarrow t = \dfrac{\ln\left(\dfrac{1}{3}\right)}{-0.02} = 54.931$$

7. $7.4 = -\log\left(\left[H^+\right]\right) \Rightarrow \left[H^+\right] = 10^{-7.4} = 0.0000000398$

8. $5 = 10(0.8)^t \Rightarrow 0.8^t = 0.5 \Rightarrow t\ln(0.8) = \ln(0.5) \Rightarrow t = \dfrac{\ln(0.5)}{\ln(0.8)} = 3.1$

9. Dinner will be ready at 6:30 p.m.

$165 = 325 - 265e^{-0.25t} \Rightarrow 265e^{-0.25t} = 160 \Rightarrow e^{-0.25t} = \dfrac{160}{265}$

$\Rightarrow -0.25t = \ln\left(\dfrac{160}{265}\right) \Rightarrow t = \dfrac{\ln\left(\dfrac{160}{265}\right)}{-0.25} = 2.01$

10. The lawn mower is 100 times louder than the garbage disposal.

Lawn mower: $100 = 10\log\left(\dfrac{I}{I_0}\right) \Rightarrow I = 10^{10}I_0;$

Garbage disposal: $80 = 10\log\left(\dfrac{I}{I_0}\right) \Rightarrow I = 10^{8}I_0$

$\dfrac{\text{Lawn mower}}{\text{Garbage disposal}} = \dfrac{10^{10}}{10^{8}} = 10^{2} = 100$

Sequences and Series

EXERCISE 7-1

1. 2.4121, 144, 167, 190, 213 (constant difference of 23)

2. J, J, A, S, O (June, July, August, September, October)

3. $1215, \dfrac{3645}{2}, \dfrac{10,935}{4}, \dfrac{32,805}{8}, \dfrac{98,415}{16}$ (constant ratio of $\dfrac{3}{2}$)

4. 70, 110, 150, 190, 230

5. $1800, -1200, 800, \dfrac{-1600}{3}, \dfrac{3200}{9}$

6. 8, 29, 92, 281, 848

7. 2; 4; 32; 4096; 536,870,912

8. 2, 4, 22, 100, 466

9. $212; a_n = 8n + 12$

10. $382, 637, 520; a_n = 80(3)^{n-1}$

11. $25,160; \dfrac{85}{2}(2 + 590)$

12. $20,438.74$ (rounded to the nearest hundredth); $\dfrac{5(1 - 1.7^{15})}{1 - 1.7}$

13. $-3800; \dfrac{50}{2}(71 - 223)$

14. $\dfrac{2,863,227,800}{177147}; \dfrac{5400\left(1 - \left(\dfrac{2}{3}\right)^{15}\right)}{1 - \dfrac{2}{3}}$

15. $70; \dfrac{49}{1 - \dfrac{3}{10}}$

EXERCISE 7-2

1. $S_{349} = \dfrac{500(1 - (1.002)^{349})}{-0.002} = \$252082.20;$

   Number of payments $= (65 - 36) * 12 + 1$

2. $P = \dfrac{35,000(1.002)^{72}(-0.002)}{1 - (1.002)^{72}} = \$522.44$

3. $P = \dfrac{475{,}000(1.00375)^{300}(-0.00375)}{1-(1.00375)^{300}} = \$2640.20$

4. Step 1: Prove for $n = 1$: $1 = \dfrac{1(2)(3)}{6}$ True.

Step 2: Assume for $n = k$: $1 + 2^2 + 3^2 + \cdots + k^2 = \dfrac{k(k+1)(2k+1)}{6}$

Step 3: Prove for $n = k + 1$:

$$1 + 2^2 + 3^2 + \cdots + k^2 + (k+1)^2 = \dfrac{(k+1)((k+1)+1)(2(k+1)+1)}{6}$$

$$= \dfrac{(k+1)(k+2)(2k+3)}{6}$$

Replace sum of first k terms with value from assumption:

$$1 + 2^2 + 3^2 + \cdots + k^2 + (k+1)^2 = \dfrac{k(k+1)(2k+1)}{6} + (k+1)^2$$

Get a common denominator:

$$\dfrac{k(k+1)(2k+1)}{6} + (k+1)^2 \Rightarrow \dfrac{k(k+1)(2k+1) + 6(k+1)^2}{6}$$

Factor:

$$\dfrac{k(k+1)(2k+1) + 6(k+1)^2}{6} \Rightarrow \dfrac{(k+1)(k(2k+1) + 6(k+1))}{6}$$

Simplify:

$$\frac{(k+1)(k(2k+1)+6(k+1))}{6} \Rightarrow \frac{(k+1)(2k^2+7k+6)}{6}$$

$$\Rightarrow \frac{(k+1)(k+2)(2k+3)}{6}$$

The proof is done.

5. $\displaystyle\sum_{n=1}^{250}(2n-1) = 250^2 = 62{,}500$; sum of the first 250 odd positive integers.

6. $\displaystyle\sum_{n=1}^{150}(2n^2) = 2\sum_{n=1}^{150}(n^2) = \frac{2(150)(151)(2(150)+1)}{6} = 2{,}272{,}550$; the sum
of the squares of the first 150 positive integers.

7. $\displaystyle\sum_{n=1}^{20}(n^3) = \left(\frac{20(21)}{2}\right)^2 = 44{,}100$; the sum of the cubes of the first 20
positive integers.

8. $\displaystyle\sum_{n=1}^{20}(n^3+n^2-n) = \sum_{n=1}^{20}(n^3) + \sum_{n=1}^{20}(n^2) - \sum_{n=1}^{20}(n)$

$$= 44{,}100 + \frac{20(21)(41)}{6} - \frac{20(21)}{2} = 46{,}760.$$

9. $\dfrac{625}{4}; \dfrac{125}{1-\dfrac{1}{5}}$

10. $\dfrac{625}{6}; \dfrac{125}{1-\dfrac{-1}{5}}$

Systems of Equations and Matrices

EXERCISE 8-1

1. $(7, 5)$

2. $\begin{bmatrix} 34 \\ 18 \end{bmatrix}$

   $\begin{bmatrix} 4(10) + 9(2) - 3(8) \\ 2(10) - 5(2) + 1(8) \end{bmatrix}$

3. $(8, -9)$

   $D = 23; Dx = -184; Dy = 207$

4. $(4, 7, 5)$

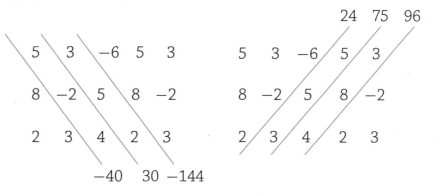

   $D = (-40 + 30 - 144) - (24 + 75 + 96) = -349$

$$588 \quad 165 \quad 516$$

$$
\begin{array}{ccc|cc}
11 & 3 & -6 & 11 & 3 \\
43 & -2 & 5 & 43 & -2 \\
49 & 3 & 4 & 49 & 3
\end{array}
\qquad
\begin{array}{ccc|cc}
11 & 3 & -6 & 11 & 3 \\
43 & -2 & 5 & 43 & -2 \\
49 & 3 & 4 & 49 & 3
\end{array}
$$

$$-88 \quad 735 \quad -774$$

$$D_x = (-88 + 735 - 774) - (588 + 165 + 516) = -1396$$

$$x = \frac{-1396}{-349} = 4$$

$$-516 \quad 1225 \quad 352$$

$$
\begin{array}{ccc|cc}
5 & 11 & -6 & 5 & 11 \\
8 & 43 & 5 & 8 & 43 \\
2 & 49 & 4 & 2 & 49
\end{array}
\qquad
\begin{array}{ccc|cc}
5 & 11 & -6 & 5 & 11 \\
8 & 43 & 5 & 8 & 43 \\
2 & 49 & 4 & 2 & 49
\end{array}
$$

$$860 \quad 110 \quad -2352$$

$$D_y = (860 + 110 - 2352) - (-516 + 1225 + 352) = -2443$$

$$y = \frac{-2443}{-349} = 7$$

$$5(4) + 3(7) - 6z = 11 \Rightarrow z = 5$$

5. $(8, -9)$

$$\begin{bmatrix} 3 & 2 \\ 4 & -5 \end{bmatrix}^{-1} \begin{bmatrix} 6 \\ 77 \end{bmatrix} = \begin{bmatrix} 8 \\ -9 \end{bmatrix}$$

6. $(4, 7, 5)$

$$\begin{bmatrix} 5 & 3 & -6 \\ 8 & -2 & 5 \\ 2 & 3 & 4 \end{bmatrix}^{-1} \begin{bmatrix} 11 \\ 43 \\ 49 \end{bmatrix} = \begin{bmatrix} 4 \\ 7 \\ 5 \end{bmatrix}$$

7. $(-9, -5, 10, 12)$

$$\begin{bmatrix} 4 & -6 & 3 & 2 \\ 8 & 2 & 5 & -7 \\ -5 & -8 & -4 & 8 \\ 20 & 13 & 19 & 25 \end{bmatrix}^{-1} \begin{bmatrix} 48 \\ -116 \\ 141 \\ 245 \end{bmatrix} = \begin{bmatrix} -9 \\ -5 \\ 10 \\ 12 \end{bmatrix}$$

8. $\left(\dfrac{-3}{2}, \dfrac{2}{3}, \dfrac{-3}{4}, \dfrac{5}{6} \right)$

$$\begin{bmatrix} 8 & 9 & -12 & 12 \\ 6 & -3 & 4 & 6 \\ 10 & 9 & -4 & -6 \\ 2 & 15 & 8 & 18 \end{bmatrix}^{-1} \begin{bmatrix} 13 \\ -9 \\ 111 \\ 16 \end{bmatrix} = \begin{bmatrix} \dfrac{-3}{2} \\ \dfrac{2}{3} \\ \dfrac{-3}{4} \\ \dfrac{5}{6} \end{bmatrix}$$

9. $\left(\dfrac{3}{2}, 16\right)$

$$\begin{aligned} 9a + 3b + c &= 7 \\ 4a - 2b + c &= -33 \\ 25a - 5b + c &= -153 \end{aligned} \Rightarrow \begin{bmatrix} 9 & 3 & 1 \\ 4 & -2 & 1 \\ 25 & -5 & 1 \end{bmatrix}^{-1} \begin{bmatrix} 7 \\ -33 \\ -153 \end{bmatrix} = \begin{bmatrix} -4 \\ 12 \\ 7 \end{bmatrix}$$

$$\Rightarrow f(x) = -4x^2 + 12x + 7$$

10. *a* is worth 2 points, *e* is worth 3 points, *l* is worth 1 point, and *v* is worth 6 points.

$$\begin{aligned} 3v + a + 2l + 3 &= 25 \\ 3v + e + 2a + l &= 26 \\ 3l + a + 2v + e &= 20 \\ 3e + v + 2a + l &= 20 \end{aligned} \quad \text{becomes} \quad \begin{aligned} 3v + a + 2l + e &= 25 \\ 3v + 2a + l + e &= 26 \\ 2v + a + 3l + e &= 20 \\ v + 2a + l + 3e &= 20 \end{aligned}$$

$$\Rightarrow \begin{bmatrix} 3 & 1 & 2 & 1 \\ 3 & 2 & 1 & 1 \\ 2 & 1 & 3 & 1 \\ 1 & 2 & 1 & 3 \end{bmatrix}^{-1} \begin{bmatrix} 25 \\ 26 \\ 20 \\ 20 \end{bmatrix} = \begin{bmatrix} 6 \\ 2 \\ 1 \\ 3 \end{bmatrix}$$

EXERCISE 8-2

1. 179

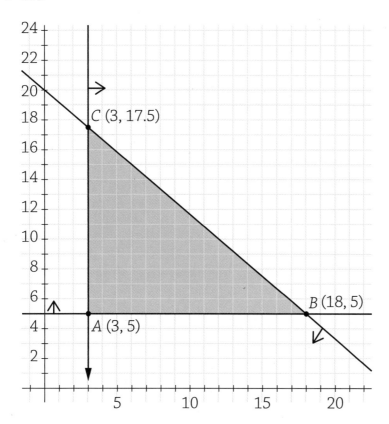

$f(A) = 8(3) + 7(5) = 59$

$f(B) = 8(18) + 7(5) = 179$

$f(C) = 8(3) + 7(17.5) = 146.5$

2. 690

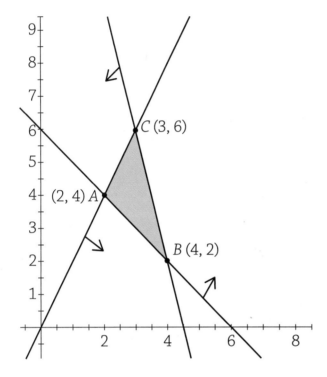

$$g(A) = 80(2) + 75(4) = 460$$
$$g(B) = 80(4) + 75(2) = 470$$
$$g(C) = 80(3) + 75(6) = 690$$

3. 450

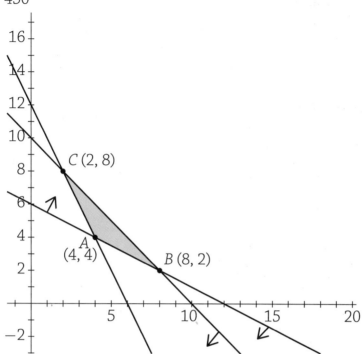

$q(A) = 30(4) + 45(4) = 300$

$q(B) = 30(8) + 45(2) = 330$

$q(C) = 30(2) + 45(8) = 450$

4. Make 75 Patty's Pig and 150 What's Your Beef for $1500 in revenue.

x: The number of Patty's Pig sandwiches made

y: The number of Patty's What's Your Beef sandwiches made

Objective function: $R(x, y) = 6x + 5y$

Restrictions: $x \geq 0$ $y \geq 0$

    Pork: $4x + 2y \leq 600$

    Beef: $4x + 4y \leq 720$

    Limits: $y \leq 2x$

Feasibility region:

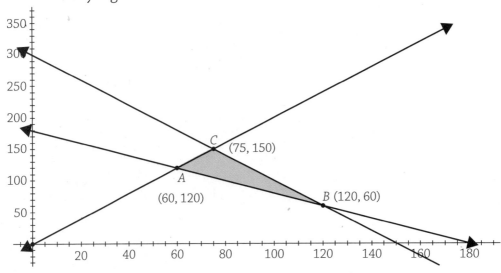

$R(A) = 6(60) + 5(120) = 960$

$R(B) = 6(120) + 5(60) = 1020$

$R(C) = 6(75) + 5(150) = 1500$

5. Make 12 dragons and 12 butterflies for a maximum profit of $132.

    x: The number of dragons made

    y: The number of butterflies made

    Objective function: $P(x,y) = 6x + 5y$

    Constraints: $x \geq 0$ $y \geq 0$

        Paper: $x + y \leq 24$

        Time: $6x + 4y \leq 120$

        Limit: $x \leq 10$

Feasibility region:

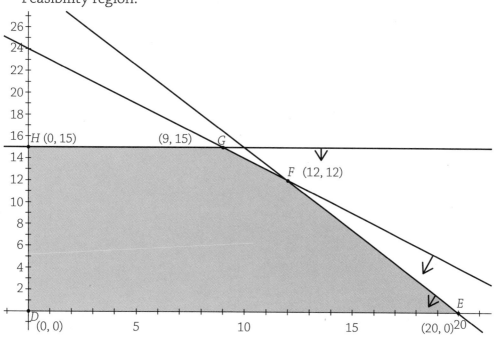

$$P(D) = 6(0) + 5(0) = 0$$
$$P(E) = 6(20) + 5(0) = 120$$
$$P(F) = 6(12) + 5(12) = 132$$
$$P(G) = 6(9) + 5(15) = 129$$
$$P(H) = 6(0) + 5(15) = 75$$

Triangle Trigonometry

EXERCISE 9-1

1. $x = 150\sin(36) = 88.2$

2. $x = 180\cos(26) = 161.8$

3. $x = 180\tan(51) = 222.3$

EXERCISE 9-2

1. $\theta = \sin^{-1}\left(\dfrac{94}{150}\right) = 38.8°$

2. $\theta = \cos^{-1}\left(\dfrac{49}{180}\right) = 74.2°$

3. $\theta = \tan^{-1}\left(\dfrac{210}{180}\right) = 49.4°$

EXERCISE 9-3

1. 468 ft.

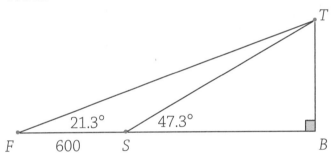

Find ST using $m \sphericalangle FTS = 20°$.

$$\frac{ST}{\sin(21.3)} = \frac{600}{\sin(20)} \Rightarrow ST = \frac{600\sin(21.3)}{\sin(20)}$$

Use the right triangle:

$$\sin(47.3) = \frac{BT}{ST} \Rightarrow BT = ST\sin(47.3) \Rightarrow BT = \frac{600\sin(21.3)\sin(47.3)}{\sin(20)}$$

2. $\dfrac{1}{2}(40)(50)\sin(49) = 755$

3. $\dfrac{1}{2}(120)(130)\sin(141) = 4909$

4. $8693 = \dfrac{1}{2}(140)(160)\sin(K) \Rightarrow K = \sin^{-1}\left(\dfrac{2(8693)}{(140)(160)}\right) = 51°$

5. $138°$

$$\sin^{-1}\left(\dfrac{2 * 71365}{520 * 410}\right) = 42.0258° \Rightarrow \theta = 180 - 42.0258$$

6. $37.5°$

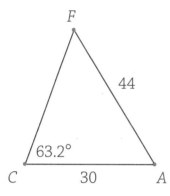

$$\dfrac{\sin(F)}{30} = \dfrac{\sin(63.2)}{44} \Rightarrow m\sphericalangle F = \sin^{-1}\left(\dfrac{30\sin(63.2)}{44}\right)$$

7. $37.3°$

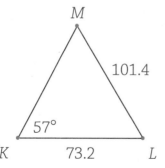

$$\dfrac{\sin(M)}{73.2} = \dfrac{\sin(57)}{101.4} \Rightarrow m\sphericalangle M = \sin^{-1}\left(\dfrac{73.2\sin(57)}{101.4}\right)$$

8. $QS = 12.0, QR = 10.1$

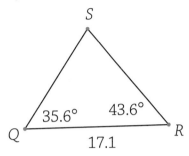

$m∢S = 100.8$

$$\frac{QS}{\sin(43.6)} = \frac{17.1}{\sin(100.8)} \Rightarrow QS = \frac{17.1\sin(43.6)}{\sin(100.8)}$$

$$\frac{QR}{\sin(35.6)} = \frac{17.1}{\sin(100.8)} \Rightarrow QR = \frac{17.1\sin(35.6)}{\sin(100.8)}$$

9. $ES = 58.7, EY = 67.3$

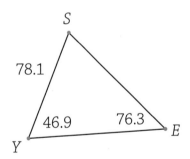

$m∢S = 56.8°$

$$\frac{ES}{\sin(46.9)} = \frac{78.1}{\sin(76.3)} \Rightarrow ES = \frac{78.1\sin(46.9)}{\sin(76.3)}$$

$$\frac{EY}{\sin(56.8)} = \frac{78.1}{\sin(76.3)} \Rightarrow EY = \frac{78.1\sin(56.8)}{\sin(76.3)}$$

EXERCISE 9-4

1. Zero

   $78\sin(40) = 50.1$ is the shortest length for which a triangle can be made; since KT is less than this number, zero triangles can be constructed.

2. Two triangles

   The shortest length of a side for which a triangle can be constructed is $128\sin(70) = 120.2$. Since YD is between this number and 128, two triangles can be constructed.

3. One triangle

   Since $TD > TA$, only one triangle can be constructed.

4. One triangle

   Since $\sphericalangle E$ is an obtuse angle and $TE > EN$, one triangle can be constructed.

EXERCISE 9-5

1. 46

   $$EC^2 = 30^2 + 45^2 - 2(30)(45)\cos(73) \Rightarrow EC^2 = 2135.6$$

2. 45.1°

   $$54^2 = 76^2 + 58^2 - 2(76)(58)\cos(X) \Rightarrow \cos(X) = \frac{54^2 - 76^2 - 58^2}{2(76)(58)}$$

3. 115.8°

   $$45.4^2 = 37.6^2 + 13.9^2 - 2(37.6)(13.9)\cos(T)$$

   $$\Rightarrow \cos(T) = \frac{45.4^2 - 37.6^2 - 13.9^2}{2(37.6)(13.9)}$$

EXERCISE 9-6

1. 129.7

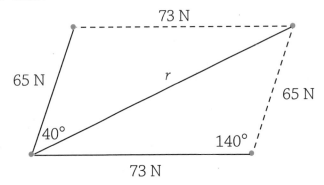

$$r^2 = 65^2 + 73^2 - 2(65)(73)\cos(140) = 16823.8$$

2. 80.0°

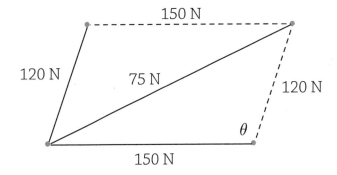

$$175^2 = 150^2 + 120^2 - 2(150)(120)\cos(\theta)$$

$$\Rightarrow \theta = \cos^{-1}\left(\frac{175^2 - 150^2 - 120^2}{-2(150)(120)}\right)$$

3. 52.3°

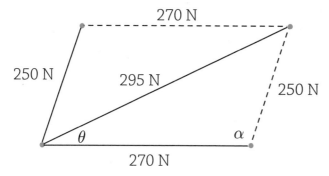

270 N

250 N

295 N

250 N

θ

α

270 N

$$295^2 = 250^2 + 270^2 - 2(250)(270)\cos(\alpha) \Rightarrow$$

$$\alpha = \cos^{-1}\left(\frac{295^2 - 250^2 - 270^2}{-2(250)(270)}\right) = 69.0$$

$$\frac{\sin(\theta)}{250} = \frac{\sin(\alpha)}{295} \Rightarrow \theta = \sin^{-1}\left(\frac{250\sin(\alpha)}{295}\right)$$

10
Trigonometric Functions

EXERCISE 10-1

1. $\dfrac{144\pi}{180} = \dfrac{4\pi}{5}$

2. $\dfrac{135\pi}{180} = \dfrac{3\pi}{4}$

3. $\dfrac{310\pi}{180} = \dfrac{31\pi}{18}$

EXERCISE 10-2

1. $\dfrac{5\pi}{8} \times \dfrac{180°}{\pi} = 112.5°$

2. $\dfrac{11\pi}{18} \times \dfrac{180°}{\pi} = 110°$

3. $\dfrac{7\pi}{12} \times \dfrac{180°}{\pi} = 105°$

EXERCISE 10-3

1. $\sec(Q) = \dfrac{17}{15}$

2. $\sin(Q) = \dfrac{8}{17}$

3. $\cot(Q) = \dfrac{15}{8}$

4. $\cos(W) = \dfrac{5}{7}$

5. $\sin(W) = \dfrac{2\sqrt{6}}{7}$

6. $\tan(W) = \dfrac{2\sqrt{6}}{5}$

EXERCISE 10-4

1. 2.5

   $20\theta = 50$

2. $\dfrac{20}{\pi}$

   $r\left(\dfrac{6\pi}{5}\right) = 20$

3. 144π

$$\frac{1}{2}(36)^2\left(\frac{2\pi}{9}\right)$$

4. $r = 40$ and $\theta = 4$ or $r = 80$ and $\theta = 2$

$2r + r\theta = 240$

$\dfrac{1}{2}r^2\theta = 3200$ yields the equation $r^2 - 120r + 3200 = 0$.

EXERCISE 10-5

1. $180° - 105° = 75°$

2. $195° - 180° = 15°$

3. $265° - 180° = 85°$

4. $540° - 517° = 23°$

EXERCISE 10-6

1. $\tan(37°)$

2. $-\sin(46°)$

3. $-\cos(42°)$

EXERCISE 10-7

1. $\cos(\theta) = \dfrac{-8}{17}$

2. $\tan(\beta) = \dfrac{-7}{24}$

3. $\csc(\alpha) = \dfrac{\sqrt{34}}{3}$

EXERCISE 10-8

1. $\sin(\alpha + \beta) = \left(\dfrac{5}{13}\right)\left(\dfrac{-4}{5}\right) + \left(\dfrac{-12}{13}\right)\left(\dfrac{-3}{5}\right) = \dfrac{16}{65}$.

2. $\cos(\alpha + \beta) = \left(\dfrac{-12}{13}\right)\left(\dfrac{-4}{5}\right) - \left(\dfrac{5}{13}\right)\left(\dfrac{-3}{5}\right) = \dfrac{63}{65}$

3. $\tan(\alpha + \beta) = \dfrac{\sin(\alpha + \beta)}{\cos(\alpha + \beta)} = \dfrac{16}{63}$

4. $\sec(\alpha - \beta) = \dfrac{1}{\cos(\alpha - \beta)} = \dfrac{65}{33}$

5. $\cos(2\beta) = \left(\dfrac{-4}{5}\right)^2 - \left(\dfrac{-3}{5}\right)^2 = \dfrac{7}{25}$

6. $\cos\left(\dfrac{\beta}{2}\right) = -\sqrt{\dfrac{1 + \dfrac{-4}{5}}{2}} = -\sqrt{\dfrac{1}{10}} = \dfrac{-\sqrt{10}}{10}$.

7. $\sin\left(\alpha + \dfrac{\pi}{3}\right) = \left(\dfrac{5}{13}\right)\left(\dfrac{1}{2}\right) + \left(\dfrac{-12}{13}\right)\left(\dfrac{\sqrt{3}}{2}\right) = \dfrac{5 - 12\sqrt{3}}{26}$

8. $\sin(2(\alpha + \beta)) = 2\sin(\alpha + \beta)\cos(\alpha + \beta) = 2\left(\dfrac{16}{65}\right)\left(\dfrac{63}{65}\right) = \dfrac{2016}{4225}$.

9. $\tan\left(\dfrac{\alpha}{2}\right) = \sqrt{\dfrac{1 - \dfrac{-12}{13}}{\dfrac{-12}{13} + 1}} = \sqrt{\dfrac{\dfrac{25}{13}}{\dfrac{1}{13}}} = \sqrt{\dfrac{25}{1}} = 5$

EXERCISE 10-9

1. $(1 - \sin^2(\theta))(1 + \tan^2(\theta)) = 1$

$(\cos^2(\theta))(\sec^2(\theta)) = 1$

$1 = 1$

2. $\tan(\theta) + \dfrac{\cos(\theta)}{1 + \sin(\theta)} = \sec(\theta)$

$\dfrac{\sin(\theta)}{\cos(\theta)} + \dfrac{\cos(\theta)}{1 + \sin(\theta)} = \dfrac{1}{\cos(\theta)}$

$\dfrac{\sin(\theta)(1 + \sin(\theta)) + \cos^2(\theta)}{\cos(\theta)(1 + \sin(\theta))} = \dfrac{1}{\cos(\theta)}$

$\dfrac{\sin(\theta) + \sin^2(\theta) + \cos^2(\theta)}{\cos(\theta)(1 + \sin(\theta))} = \dfrac{1}{\cos(\theta)}$

$\dfrac{\sin(\theta) + 1}{\cos(\theta)(1 + \sin(\theta))} = \dfrac{1}{\cos(\theta)}$

$\dfrac{1}{\cos(\theta)} = \dfrac{1}{\cos(\theta)}$

3. $\dfrac{\sec(\beta) - 1}{\sec(\beta) + 1} = \dfrac{1 - \cos(\beta)}{1 + \cos(\beta)}$

$\dfrac{\dfrac{1}{\cos(\beta)} - 1}{\dfrac{1}{\cos(\beta)} + 1} = \dfrac{1 - \cos(\beta)}{1 + \cos(\beta)}$

$\left(\dfrac{\dfrac{1}{\cos(\beta)} - 1}{\dfrac{1}{\cos(\beta)} + 1}\right) \dfrac{\cos(\beta)}{\cos(\beta)} = \dfrac{1 - \cos(\beta)}{1 + \cos(\beta)}$

$$\frac{1 - \cos(\beta)}{\cos(\beta) + 1} = \frac{1 - \cos(\beta)}{1 + \cos(\beta)}$$

4. $\dfrac{\tan(\alpha) + \tan(\beta)}{\cot(\alpha) + \cot(\beta)} = \tan(\alpha)\tan(\beta)$

This is one of the few problems that we will not convert to $\sin(\theta)$ and $\cos(\theta)$ but will leave in terms of $\tan(\theta)$.

$$\frac{\tan(\alpha) + \tan(\beta)}{\dfrac{1}{\tan(\alpha)} + \dfrac{1}{\tan(\beta)}} = \tan(\alpha)\tan(\beta)$$

$$\left(\frac{\tan(\alpha) + \tan(\beta)}{\dfrac{1}{\tan(\alpha)} + \dfrac{1}{\tan(\beta)}}\right)\frac{\tan(\alpha)\tan(\beta)}{\tan(\alpha)\tan(\beta)} = \tan(\alpha)\tan(\beta)$$

$$\left(\frac{\tan(\alpha) + \tan(\beta)}{\tan(\beta) + \tan(\alpha)}\right)\tan(\alpha)\tan(\beta) = \tan(\alpha)\tan(\beta)$$

$$\tan(\alpha)\tan(\beta) = \tan(\alpha)\tan(\beta)$$

5. $\sin(4x) = \cos(x)(4\sin(x) - 8\sin^3(x))$

$\sin(4x) = 4\sin(x)\cos(x)(1 - 2\sin^2(x))$

$\sin(4x) = 2(2\sin(x)\cos(x))(\cos(2x))$

$\sin(4x) = 2\sin(2x)\cos(2x)$

$\sin(4x) = \sin(4x)$

11
Graphs and Applications of Trigonometric Functions

EXERCISE 11-1

1. Up 1; key points: $(0, 1)$, $\left(\dfrac{\pi}{2}, 2\right)$, $(\pi, 1)$, $\left(\dfrac{3\pi}{2}, 0\right)$, $(2\pi, 1)$

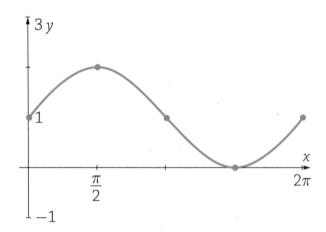

2. Up 1; key points: $(0, 2)$, $\left(\dfrac{\pi}{2}, 1\right)$, $(\pi, 0)$, $\left(\dfrac{3\pi}{2}, 1\right)$, $(2\pi, 2)$

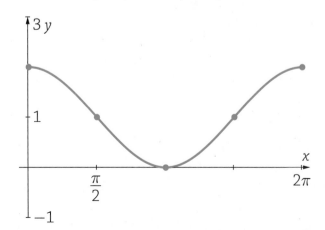

3. Stretch from the x-axis by a factor of 2; key points:

$$(0,0), \left(\frac{\pi}{2}, 2\right), (\pi, 0), \left(\frac{3\pi}{2}, -2\right), (2\pi, 0)$$

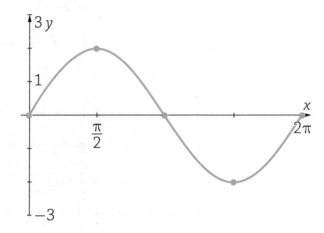

4. Dilate to y-axis by a factor of 2; key points:

$$(0,0), \left(\frac{\pi}{4}, 1\right), \left(\frac{\pi}{2}, 0\right), \left(\frac{3\pi}{4}, -1\right), (\pi, 0)$$

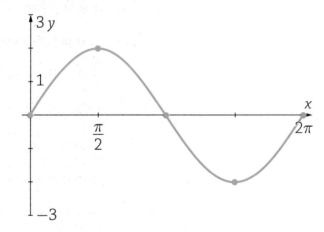

5. Dilate to y-axis by a factor of π; key points:

$$(0,1), \left(\frac{1}{2}, 0\right), (1, -1), \left(\frac{3}{2}, 0\right), (2, 1)$$

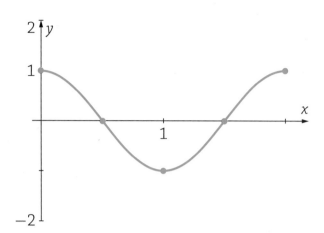

6. Dilate to y-axis by a factor of 2, dilate from x-axis by a factor of 2, down 1; key points: $(0,1), \left(\dfrac{\pi}{4}, -1\right), \left(\dfrac{\pi}{2}, -3\right), \left(\dfrac{3\pi}{4}, -1\right), (\pi, 1)$

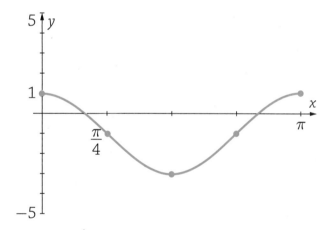

7. Right $\dfrac{\pi}{2}$, reflect over the x-axis, dilate from the x-axis by a factor of $\dfrac{1}{2}$; key points: $\left(\dfrac{\pi}{2}, 0\right), \left(\pi, \dfrac{-1}{2}\right), \left(\dfrac{3\pi}{2}, 0\right), \left(2\pi, \dfrac{1}{2}\right), \left(\dfrac{5\pi}{2}, 0\right)$

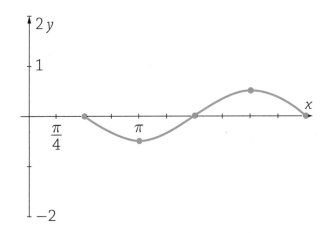

8. Left $\dfrac{\pi}{4}$, dilate from x-axis by a factor of 2, up 1; key points:

$$\left(\dfrac{-\pi}{4},1\right),\left(\dfrac{\pi}{4},3\right),\left(\dfrac{3\pi}{4},1\right),\left(\dfrac{5\pi}{4},-1\right),\left(\dfrac{7\pi}{4},1\right)$$

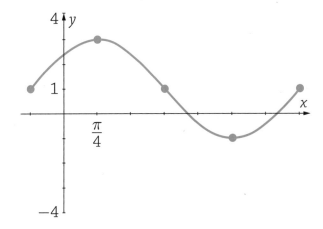

9. Left $\dfrac{\pi}{3}$, reflect over x-axis, dilate from x-axis by a factor of 2, up 3;

key points: $\left(\dfrac{-\pi}{3},1\right),\left(\dfrac{\pi}{6},3\right),\left(\dfrac{2\pi}{3},5\right),\left(\dfrac{7\pi}{6},3\right),\left(\dfrac{5\pi}{3},1\right)$

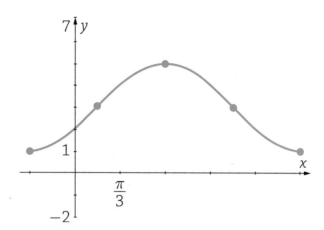

10. h; the period of the function is 2π, so the point $(z + 2\pi, h)$ lies one cycle to the left of (z, h).

EXERCISE 11-2

1. $y = 3\cos(2x) + 1$

   Amplitude $= \dfrac{4 - (-2)}{2} = 3$; average value $= \dfrac{4 + (-2)}{2} = 1$; period is π.

2. $y = 2\sin\left(\dfrac{1}{2}\left(x - \dfrac{\pi}{4}\right)\right) - 1$ or $y = 2\cos\left(\dfrac{1}{2}\left(x - \dfrac{5\pi}{4}\right)\right) - 1$

   Amplitude $= \dfrac{1 - (-3)}{2} = 2$; average value $= \dfrac{1 + (-3)}{2} = -1$; period is 4π.

3. $y = -4\sin\left(\dfrac{\pi}{3}x\right) + 2$

   Amplitude $= \dfrac{6 - (-2)}{2} = 4$; average value $= \dfrac{6 + (-2)}{2} = 2$; period is 6.

4. $y = -4\cos\left(3\left(x + \dfrac{\pi}{6}\right)\right) + 5$

   Amplitude $= \dfrac{9 - (1)}{2} = 4$; average value $= \dfrac{9 + (1)}{2} = 5$; period is $\dfrac{2\pi}{3}$.

5. $y = 2\tan\left(\dfrac{\pi}{4}x\right)$

Passes through (2, 1); period is 4.

EXERCISE 11-3

1. $\dfrac{-\sqrt{5}}{3}$

If $\sin(A) = \dfrac{-\sqrt{5}}{4}$, $\cos(A) = \dfrac{3}{4}$.

2. $\dfrac{-5}{\sqrt{74}}$

If $\tan(A) = \dfrac{-5}{7}$, the hypotenuse of the right triangle has length $\sqrt{74}$.

3. $\dfrac{42}{58}$

If $\tan(A) = \dfrac{3}{7}$, the hypotenuse of the right triangle has length $\sqrt{58}$;

$\sin(A) = \dfrac{3}{\sqrt{58}}$; $\cos(A) = \dfrac{7}{\sqrt{58}}$

4. $\dfrac{5\sqrt{3} + 2}{12}$

If $\sin(A) = \dfrac{2}{3}$ then $\cos(A) = \dfrac{\sqrt{5}}{3}$. If $\sin(B) = \dfrac{-1}{4}$ then $\cos(B) = \dfrac{\sqrt{15}}{4}$.

$\cos(A + B) = \left(\dfrac{\sqrt{5}}{3}\right)\left(\dfrac{\sqrt{15}}{4}\right) - \left(\dfrac{2}{3}\right)\left(\dfrac{-1}{4}\right)$

5. $\dfrac{18}{x^2+9} - 1 = \dfrac{18-(x^2+9)}{x^2+9} = \dfrac{9-x^2}{x^2+9}$

If $\tan(A) = \dfrac{x}{3}$, the hypotenuse of the right triangle has length $\sqrt{x^2+9}$

and $\cos(A) = \dfrac{3}{\sqrt{x^2+9}}$.

$\cos(2A) = 2\left(\dfrac{3}{\sqrt{x^2+9}}\right)^2 - 1$

EXERCISE 11-4

1. 228.6°, 312.4°

   Reference angle $= \sin^{-1}\left(\dfrac{3}{4}\right) = 48.6°$

2. 60.9°, 240.9°

   Reference angle $= \tan^{-1}\left(\dfrac{9}{5}\right) = 60.9°$

3. 0°, 60°, 300°

   $(2\cos(A) - 1)\,(\cos(A) - 1) = 0 \Rightarrow \cos(A) = \dfrac{1}{2},\ 1$

4. 19.5°, 160.5°

   $(3\sin(A) - 1)\,(\sin(A) + 2) = 0 \Rightarrow \sin(A) = \dfrac{1}{3},\ -2;$ reject -2

   Reference angle: $\sin^{-1}\left(\dfrac{1}{3}\right) = 19.5°$

5. 16.6°, 124.8°, 196.6°, 304.8°

   $\tan(A) = \dfrac{-8 \pm \sqrt{8^2 - 4(7)(-3)}}{2(7)} = \dfrac{-8 \pm \sqrt{148}}{14} = 0.2975,\ -1.4404$

Reference angle 1: $\tan^{-1}(0.2975) = 16.6°$; reference angle 2:

$\tan^{-1}(1.4404) = 55.2°$

6. $0°, 60°, 180°, 300°$

$2\sin(A)\cos(A) - \sin(A) = 0 \Rightarrow \sin(A)(2\cos(A) - 1) = 0$

$\Rightarrow \sin(A) = 0 \text{ or } \cos(A) = \dfrac{1}{2}$

7. $60°, 90°, 270°, 300°$

$2\cos^2(A) - 1 + \cos(A) + 1 = 0 \Rightarrow \cos(A)(2\cos(A) + 1) = 0$

$\Rightarrow \cos(A) = 0, \dfrac{1}{2}$

8. $23°, 157°, 218.8°, 320.2°$

$2(1 - 2\sin^2(A)) - \sin(A) - 1 = 0 \Rightarrow -4\sin^2(A) - \sin(A) + 1 = 0$

$\sin(A) = \dfrac{1 \pm \sqrt{(1)^2 - 4(-4)(1)}}{-8} \Rightarrow \sin(A) = \dfrac{1 \pm \sqrt{17}}{-8} = -0.640, 0.390$

Reference angle 1: $\sin^{-1}(0.640) = 39.8°$; reference angle 2:

$\sin^{-1}(0.390) = 23.0°$

9. $60°, 300°$

$4(2\cos^2(A) - 1) - \dfrac{1}{\cos(A)} + 4 = 0$

$\Rightarrow 8\cos^2(A) - \dfrac{1}{\cos(A)} = 0 \Rightarrow 8\cos^3(A) - 1 = 0$

$\Rightarrow (2\cos(A) - 1)(4\cos^2(A) + 2\cos(A) + 1) = 0 \Rightarrow \cos(A) = \dfrac{1}{2}$

10. $0°, 45°, 135°, 180°, 225°, 315°$

$2\sin(A)\cos(A) = \dfrac{\sin(A)}{\cos(A)} \Rightarrow 2\sin(A)\cos^2(A) - \sin(A) = 0$

$\Rightarrow \sin(A)(2\cos^2(A) - 1) = 0 \Rightarrow \sin(A) = 0 \text{ or } \cos(A) = \pm\dfrac{\sqrt{2}}{2}$

EXERCISE 11-5

1. 0.73, 2.41

   Reference angle: $\sin^{-1}\left(\dfrac{2}{3}\right) = .7297$

2. $\dfrac{\pi}{6}, \dfrac{\pi}{3}, \dfrac{2\pi}{3}, \dfrac{5\pi}{6}, \dfrac{7\pi}{6}, \dfrac{4\pi}{3}, \dfrac{5\pi}{3}, \dfrac{11\pi}{6}$

   $\sin^2(2\theta) = \dfrac{3}{4} \Rightarrow \sin(2\theta) = \pm\dfrac{\sqrt{3}}{2}$

   $\Rightarrow 2\theta = \dfrac{\pi}{3}, \dfrac{2\pi}{3}, \dfrac{4\pi}{3}, \dfrac{5\pi}{3}, \dfrac{7\pi}{3}, \dfrac{8\pi}{3}, \dfrac{10\pi}{3}, \dfrac{11\pi}{3}$

3. 0.45, 2.69 $(\pi - A)$, 4.02 $(\pi + B)$, 5.41 $(2\pi - B)$

   $\sin(\theta) = \dfrac{-1 \pm \sqrt{1 - 4(-1)(3)}}{6} = \dfrac{-1 \pm \sqrt{13}}{6} = 0.4343, -0.7676$

   Reference angle A: $\sin^{-1}(0.4343) = 0.449$; Reference angle B:

   $\sin^{-1}(0.7676) = 0.875$

4. 0.72, 5.56 $(2\pi - \text{reference angle})$

   $4\cos(\theta) = \dfrac{6}{\cos(\theta)} - 5 \Rightarrow 4\cos^2(\theta) + 5\cos(\theta) - 6 = 0$

   $\Rightarrow (4\cos(\theta) - 3)(\cos(\theta) + 2) = 0$

   $\Rightarrow \cos(\theta) = \dfrac{3}{4}, -2$. Reject -2 as a value for cosine. Reference angle:

   $\cos^{-1}\left(\dfrac{3}{4}\right) = 0.723$

5. $\dfrac{\pi}{6}, \dfrac{3\pi}{2}, \dfrac{5\pi}{6}$

   $\dfrac{1}{\sin(\theta)} = 2\sin(\theta) + 1 \Rightarrow 2\sin^2(\theta) + \sin(\theta) - 1 = 0$

   $\Rightarrow (2\sin(\theta) - 1)(\sin(\theta) + 1) = 0 \Rightarrow \sin(\theta) = \dfrac{1}{2}, -1$

EXERCISE 11-6

1. 365

2. $\dfrac{89 - 41}{2} = 24$

3. $\dfrac{89 + 41}{2} = 65$

4. $h(t) = -24\cos\left(\dfrac{2\pi}{365}t\right) + 65$

5. $h(100) = -24\cos\left(\dfrac{200\pi}{365}\right) + 65 = 68.6$

12
Polar and Parametric Equations

EXERCISE 12-1

1. $(-5, 275°)$

2. $(-6, 35°)$

3. $\left(-7, \dfrac{\pi}{6}\right)$

4. $\left(-8, \dfrac{\pi}{9}\right)$

EXERCISE 12-2

1. $(6, 45°)$

$r = \sqrt{(3\sqrt{2})^2 + (3\sqrt{2})^2} = 6; \theta = \tan^{-1}(1) = 45°$

2. $\left(10, \dfrac{7\pi}{6}\right)$

$$r = \sqrt{(5\sqrt{3})^2 + (5)^2} = 10; \text{Reference angle: } \tan^{-1}\left(\frac{\sqrt{3}}{3}\right) = \frac{\pi}{6}$$

3. $(\sqrt{41}, 2.246)$

$$r = \sqrt{(-4)^2 + (5)^2} = \sqrt{41}; \text{Reference angle: } \tan^{-1}\left(\frac{5}{4}\right) = 0.8961$$

EXERCISE 12-3

1. $(-6\sqrt{3}, -6)$

$x = 12\cos(210°); y = 12\sin(210°)$

2. $(-5\sqrt{2}, 5\sqrt{2})$

$$x = 10\cos\left(\frac{3\pi}{4}\right); y = 10\sin\left(\frac{3\pi}{4}\right)$$

3. $(6.180, -19.021)$

$$x = -20\cos\left(\frac{3\pi}{5}\right); y = -20\sin\left(\frac{3\pi}{5}\right)$$

EXERCISE 12-4

1. $(4, 0), \left(4, \frac{2\pi}{3}\right), \left(4, \frac{4\pi}{3}\right)$

Tips are $\frac{2\pi}{3}$ radians apart.

2. $\left(8, \frac{\pi}{10}\right), \left(8, \frac{\pi}{2}\right), \left(8, \frac{9\pi}{10}\right), \left(8, \frac{13\pi}{10}\right), \left(8, \frac{17\pi}{10}\right)$

First tip at $5\theta = \frac{\pi}{2}$ or $\theta = \frac{\pi}{10}$. Tips are $\frac{2\pi}{5}$ radians apart.

3. $(10, 0), \left(10, \frac{\pi}{6}\right), \left(10, \frac{\pi}{3}\right), \left(10, \frac{\pi}{2}\right), \left(10, \frac{2\pi}{3}\right), \left(10, \frac{5\pi}{6}\right), (10, \pi), \left(10, \frac{7\pi}{6}\right),$

$\left(10, \frac{4\pi}{3}\right), \left(10, \frac{3\pi}{2}\right), \left(10, \frac{5\pi}{3}\right), \left(10, \frac{11\pi}{6}\right)$

There are 12 petals. Tips are $\dfrac{\pi}{6}$ radians apart.

4. $\left(5,\dfrac{\pi}{8}\right),\left(5,\dfrac{3\pi}{8}\right),\left(5,\dfrac{5\pi}{8}\right),\left(5,\dfrac{7\pi}{8}\right),\left(5,\dfrac{9\pi}{8}\right),\left(5,\dfrac{11\pi}{8}\right),\left(5,\dfrac{13\pi}{8}\right),\left(5,\dfrac{15\pi}{8}\right)$

There 8 petals. First tip is at $\theta=\dfrac{\pi}{8}$. Tips are $\dfrac{\pi}{4}$ radians apart.

EXERCISE 12-5

1. $(10,0),\left(5,\dfrac{\pi}{2}\right),(0,\pi),\left(5,\dfrac{3\pi}{2}\right)$

2. $(4,0),\left(8,\dfrac{\pi}{2}\right),(4,\pi),\left(0,\dfrac{3\pi}{2}\right)$

3. $(0,0),\left(3,\dfrac{\pi}{2}\right),(6,\pi),\left(3,\dfrac{3\pi}{2}\right)$

4. $(6,0),\left(0,\dfrac{\pi}{2}\right),(6,\pi),\left(12,\dfrac{3\pi}{2}\right)$

EXERCISE 12-6

1. $(9,0),\left(3,\dfrac{\pi}{2}\right),(-3,\pi),\left(3,\dfrac{3\pi}{2}\right),\left(0,\dfrac{2\pi}{3}\right),\left(0,\dfrac{4\pi}{3}\right)$

2. $(2\sqrt{2},0),\left(2\sqrt{2}+4,\dfrac{\pi}{2}\right),(2\sqrt{2},\pi),\left(2\sqrt{2}-4,\dfrac{3\pi}{2}\right),\left(0,\dfrac{5\pi}{4}\right),\left(0,\dfrac{7\pi}{4}\right)$

3. $(-2,0),\left(3,\dfrac{\pi}{2}\right),(8,\pi),\left(3,\dfrac{3\pi}{2}\right),(0,0.54),(0,5.74)$

Reference angle $\tan^{-1}\left(\dfrac{3}{5}\right)=0.54$

4. $(4,0),\left(-2,\dfrac{\pi}{2}\right),(4,\pi),\left(10,\dfrac{3\pi}{2}\right),(0,0.73),(0,2.41)$

Reference angle $\sin^{-1}\left(\dfrac{2}{3}\right)=0.73$

EXERCISE 12-7

1. $r = 2\cos(3\theta)$

2. $r = 3\sin(4\theta)$

3. $r = 2 + 2\cos(\theta)$

4. $r = 3 - 3\sin(\theta)$

5. $r = 3 + 5\cos(\theta)$

6. $r = 3 + 2\sin(\theta)$

7. $r = 4 - 3\cos(\theta)$

8. $r = 2 + 4\sin(\theta)$

EXERCISE 12-8

1. Ellipse; eccentricity $= \dfrac{3}{5}$

2. Hyperbola; eccentricity $= \dfrac{5}{3}$

3. Parabola; eccentricity $= 1$

EXERCISE 12-9

1. $y = \dfrac{-1}{3}x$

   Substitute $t = \dfrac{x-3}{2}$.

2. $y = \log_2(8x^2)$

   $y = 2\log_2(x) + 3 = \log_2(x^2) + \log_2(8) = \log_2(8x^2)$

   Substitute $t = \log_2(x)$.

3. $\dfrac{x^2}{16} - \dfrac{y^2}{9} = 1$

   $\dfrac{x}{4} = \sec(t); \dfrac{y}{3} = \tan(t)$

Use the Pythagorean identity: $1 + \tan^2(t) = \sec^2(t)$

4. $\dfrac{x^2}{36} + \dfrac{y^2}{9} = 1$

EXERCISE 12-10

1. $y(t) = -16t^2 + 85.1\sin(40°)t + 50;\ x(t) = 85.1\cos(40°)t$
2. $y(1.71) = 96.8$ feet

   Axis of symmetry: $t = \dfrac{-85.1\sin(40°)}{-32} = 1.71$
3. $x(4.17) = 271.7$ feet

   The ball strikes the ground when $y(t) = 0$. This happens when $t = 4.17$ seconds.

Complex Numbers

EXERCISE 13-1

1. $12\,cis\left(\dfrac{5\pi}{3}\right)$

2. $16\,cis\left(\dfrac{5\pi}{4}\right)$

EXERCISE 13-2

1. $-6\sqrt{3} - 6i$
2. $10\sqrt{2} - 10i\sqrt{2}$

EXERCISE 13-3

1. Polar: $216\,cis\left(\dfrac{7\pi}{6}\right)$

   Rectangular: $-108\sqrt{3} - 108$

2. Polar: $80\,cis\left(\dfrac{25\pi}{12}\right) = 80\,cis\left(\dfrac{\pi}{12}\right)$

   Rectangular: $20\sqrt{6} + 20\sqrt{2} + (20\sqrt{6} - 20\sqrt{2})i$

3. Polar: $288\,cis\left(\dfrac{\pi}{3}\right)$

   Rectangular: $144 + 144i\sqrt{3}$

4. Polar: $2\,cis\left(\dfrac{\pi}{12}\right)$

   Rectangular: $3\sqrt{6} + 3\sqrt{2} + (3\sqrt{6} - 3\sqrt{2})i$

5. Polar: $3\,cis\left(\dfrac{\pi}{2}\right)$

   Rectangular: $3i$

6. Polar: $216\,cis(2\pi) = 216\,cis(0)$

   Rectangular: 216

7. Polar: $2^8\,cis(2\pi) = 2^8\,cis(0)$

   Rectangular: 256

EXERCISE 13-4

1. $z_1 = 2\,cis\left(\dfrac{\pi}{5}\right), \quad z_2 = 2\,cis\left(\dfrac{8\pi}{15}\right), \quad z_3 = 2\,cis\left(\dfrac{13\pi}{15}\right), \quad z_4 = 2\,cis\left(\dfrac{6\pi}{5}\right),$

   $z_5 = 2\,cis\left(\dfrac{23\pi}{15}\right), \quad z_4 = 2\,cis\left(\dfrac{28\pi}{15}\right)$

2. $z_1 = 2^{1/8}\,cis\left(\dfrac{\pi}{16}\right), \quad z_2 = 2^{1/8}\,cis\left(\dfrac{9\pi}{16}\right), \quad z_3 = 2^{1/8}\,cis\left(\dfrac{17\pi}{16}\right),$

   $z_4 = 2^{1/8}\,cis\left(\dfrac{25\pi}{16}\right)$

EXERCISE 13-5

1. $1 cis\left(\dfrac{\pi}{3}\right) = \dfrac{1}{2} + \dfrac{\sqrt{3}}{2} i$

2. $1 cis(\pi) = -1$

   (The result is often written as $e^{i\pi} + 1 = 0$ because it contains the five most important numbers in all of mathematics: 0, 1, i, e, and π.)

Introduction to Calculus

EXERCISE 14-1

1. 0

$$\lim_{x \to 5} \frac{(x-5)^2}{(x-5)(x+5)} = \lim_{x \to 5} \frac{x-5}{x+5} = 0$$

2. $\dfrac{1}{6}$

$$\lim_{x \to 9} \frac{\sqrt{x}-3}{\left(\sqrt{x}-3\right)\left(\sqrt{x}+3\right)} = \lim_{x \to 9} \frac{1}{\sqrt{x}+3} = \frac{1}{6}$$

3. $\dfrac{3}{4}$

$$\lim_{x \to \infty} \frac{3 - \dfrac{10}{x} + \dfrac{25}{x^2}}{4 - \dfrac{25}{x^2}} = \frac{3}{4}$$

4. 3

$$\lim_{x \to -2^-} x^2 - 1 = \lim_{x \to -2^+} 4x + 11 = 3$$

5. DNE

$$\lim_{x \to 3^-} x^2 - 1 = 8 \ \& \ \lim_{x \to 3^+} 13 - x = 10$$

6. 2

$$\lim_{2x \to 2(0)} \frac{2\tan(2x)}{2x} = 2 \lim_{Q \to 0} \frac{\tan(Q)}{Q} = 2(1) = 2$$

7. $\dfrac{-1}{2}$

$$\lim_{\Delta x \to 0} \frac{\cos\left(\frac{\pi}{6} + \Delta x\right) - \cos\left(\frac{\pi}{6}\right)}{\Delta x} = f'\left(\frac{\pi}{6}\right) \text{ when } f(x) = \cos(x)$$

EXERCISE 14-2

1. Yes, the value of the limit and the value of the function are the same.

$$\lim_{x \to 5} \frac{(x+5)(x-5)}{x-5} = 10 = f(5)$$

2. No, the value of the limit does not equal the value of the function.

$$\lim_{x \to 4} \frac{(x+4)(x-4)}{x-4} = 8 \neq f(4)$$

3. Yes, for all values of x, the limiting value is equal to the functional value.

EXERCISE 14-3

1. $g'(x) = 40x^7 - 36x^3$

2. $k'(x) = [1]\sqrt{x^2 + 3} + x\left[\dfrac{1}{2}(x^2 + 3)^{-\frac{1}{2}}(2x)\right] = \sqrt{x^2 + 3} + \dfrac{x^2}{\sqrt{x^2 + 3}}$

3. $p'(x) = \dfrac{[2](5x - 4) - (2x + 3)[5]}{(5x - 4)^2} = \dfrac{-23}{(5x - 4)^2}$

4. $t(x) = \dfrac{\sin(x)}{\cos(x)} \Rightarrow t'(x) = \dfrac{[\cos(x)]\cos(x) - \sin(x)[-\sin(x)]}{\cos^2(x)}$

$= \dfrac{\cos^2(x) + \sin^2(x)}{\cos^2(x)} = \sec^2(x)$

5. $w'(x) = 3\sin^2(x^2)\cos(x^2)[2x] = 6x\sin^2(x^2)\cos(x^2)$

6. $v'(x) = \dfrac{[[1]\cos(x) + x[-\sin(x)]](x^2 - 1) - x\cos(x)[2x]}{(x^2 - 1)^2}$

$= \dfrac{(\cos(x) - x\sin(x))(x^2 - 1) - 2x^2\cos(x)}{(x^2 - 1)^2}$

EXERCISE 14-4

1. $y - 45 = 64(x - 2)$

   $f(2) = 45; f'(x) = 15x^2 + 4; f'(2) = 64$

2. $y + 2\sqrt{3} = 4\left(x - \dfrac{5\pi}{6}\right)$

   $g\left(\dfrac{5\pi}{6}\right) = -2\sqrt{3}; g'(x) = 8\cos(2x); g'\left(\dfrac{5\pi}{6}\right) = 4$

EXERCISE 14-5

1. $(-2, -60), (0, 4), (3, -185)$

   $b'(x) = 12x^3 - 12x^2 - 72x = 12x(x^2 - x - 6) = 12x(x - 3)(x + 2)$

2. $\left(-1, \dfrac{-1}{2}\right); \left(1, \dfrac{1}{2}\right)$

   $c'(x) = \dfrac{[1](x^2 + 1) - x[2x]}{(x^2 + 1)^2} = \dfrac{1 - x^2}{(x^2 + 1)^2} = \dfrac{(1 - x)(1 + x)}{(x^2 + 1)^2}$